NIELS BOHR: HIS HERITAGE AND LEGACY

Science and Philosophy

VOLUME 6

This series has been established as a forum for contemporary analysis of philosophical problems which arise in connection with the construction of theories in the physical and the biological sciences. Contributions will not place particular emphasis on any one school of philosophical thought. However, they will reflect the belief that the philosophy of science must be firmly rooted in an examination of actual scientific practice. Thus, the volumes in this series will include or depend significantly upon an analysis of the history of science, recent or past. The Editors welcome contributions from scientists as well as from philosophers and historians of science.

The titles published in this series are listed at the end of this volume.

JAN FAYE

NIELS BOHR: HIS HERITAGE AND LEGACY

An Anti-Realist View of Quantum Mechanics

Kluwer Academic Publishers

Dordrecht / Boston / London

Library of Congress Cataloging-in-Publication Data

Faye, Jan.
Niels Bohr : his heritage and legacy : an anti-realist view of quantum mechanics / Jan Faye.
p. cm. -- (Science and philosophy ; 6)
Includes bibliographical references and index.
ISBN 0-7923-1294-5 (alk. paper)
1. Quantum theory--History. 2. Bohr, Niels Hendrik David, 1885-1962--Knowledge--Philosophy. 3. Hoffding, Harald, 1843-1931--Influence. I. Title. II. Series.
QC173.98.F38 1991
530.1'2--dc20 91-17115

ISBN 0-7923-1294-5

Published by Kluwer Academic Publishers,
P.O. Box 17, 3300 AA Dordrecht, The Netherlands.

Kluwer Academic Publishers incorporates
the publishing programmes of
D. Reidel, Martinus Nijhoff, Dr W. Junk and MTP Press.

Sold and distributed in the U.S.A. and Canada
by Kluwer Academic Publishers,
101 Philip Drive, Norwell, MA 02061, U.S.A.

In all other countries, sold and distributed
by Kluwer Academic Publishers Group,
P.O. Box 322, 3300 AH Dordrecht, The Netherlands.

Printed on acid-free paper

Printed in The Netherlands

To Mona

Table of Contents

Preface

The bulk of the present book has not been published previously though Chapters II and IV are based in part on two earlier papers of mine: "The Influence of Harald Høffding's Philosophy on Niels Bohr's Interpretation of Quantum Mechanics", which appeared in *Danish Yearbook of Philosophy*, 1979, and "The Bohr-Høffding Relationship Reconsidered", published in *Studies in History and Philosophy of Science*, 1988. These two papers complement each other, and in order to give the whole issue a more extended treatment I have sought, in the present volume by drawing on relevant historical material, to substantiate the claim that Høffding was Bohr's mentor. Besides containing a detailed account of Bohr's philosophy, the book, at the same time, serves the purpose of making Høffding's ideas and historical significance better known to a non-Danish readership.

During my work on this book I have consulted the Royal Danish Library; the National Archive of Denmark and the Niels Bohr Archive, Copenhagen, in search of relevant material. I am grateful for permission to use and quote material from these sources. Likewise, I am indebted to colleagues and friends for commenting upon the manuscript: I am especially grateful to Professor Henry Folse for our many discussions during my visit to New Orleans in November-December 1988 and again here in Elsinore in July 1990. I have benefitted from the generosity with which he has commented upon the text and his suggestions as to how the representation of my considerations might be improved, despite the fact that there are points of interpretation on which we do not agree. I would also like to thank Finn Collin for his comments on Chapter VIII.

Furthermore, I owe special thanks to Susan Dew for her efforts to correct my English prose, which she sometimes had to rewrite in the interest of style. Finally, I want to express my debt to the Carlsberg Foundation for financial support during the writing of the book.

Elsinore, August 1990

Prologue: The Heritage

On the wall above Niels Bohr's desk in his study in the Carlsberg honorary residence hang, besides a few small photographs and paintings of his mother and brother, two large paintings, one of his father and the other of the Danish philosopher and psychologist Harald Høffding.[1] Høffding preceded Bohr at Carlsberg; he had been a close friend of Niels Bohr's father, Christian Bohr, he had been the younger Bohr's teacher at the University, and later they became friends. I will attempt to show that Høffding's philosophical influence on the young Bohr was direct and exceedingly important, in spite of the fact that it has more than once been asserted that Høffding's influence on Bohr was of a more indirect nature, that it took the form of inspiring him to grasp what it was that unified all human endeavour in the search for knowledge.[2] So, if I am right, when working Bohr had before him not only the portrait of his biological father but also, I suggest, that of his intellectual father.

Without doubt, the intellectual climate in which a scientist grows up often has a considerable influence on his scientific work. Even if not immediately obvious to the working scientist himself, this climate is nevertheless of decisive significance for the molding of his beliefs and ideas, and thus also for his reaction to and comprehension of anything new he encounters. If he is also one of the leading figures in an entirely new and thriving field of science, and if he is furthermore a little older than the other scientists, it is reasonable to conjecture that his personality and authority will leave its impress on all scientific thought within this field to such an extent as to make it at times difficult to get a glimpse of the philosophical background for his work.

The intention behind the present book is that of demonstrating that there was a close connection between Niels Bohr's approach to the study of the atom and the philosophical influences which shaped his outlook from childhood and youth onwards. Bohr often spoke of "the epistemological instruction" that the latest developments in atomic physics had supplied, but this instruction, of course, cannot be ingested without a further account of the philosophy on which the interpretation of quantum mechanics is based. In the case of the interpretation of specific phenomena, which in themselves are not theoretically unambiguous, any such interpretation will normally be colored by the philosophical assumptions of the interpreter. This also applied in the case of Niels Bohr.

My claim that Høffding was Bohr's intellectual father and mentor is supported not only by bibliographical material but also by the results of an extensive analysis of their theories of epistemology. In particular, I shall draw attention to Høffding's conception of reality, his analysis of the relation between subject and object, and the attention he gave to the "complementary" conditions for description in his treatment of psychological experience and living organisms. All of these themes were of great importance to Bohr.

The day before Bohr died, on 17th November 1962, he gave an interview to Thomas S. Kuhn, Aage Petersen and Erik Rüdinger in his office at Carlsberg.[3] In this last interview Bohr tells us something of his early interest in philosophy, psychology and biology, and more than once he returns to Høffding's name. The interview is not very coherent, and Bohr seems to be marked by old age. However, we should remember that he had to express himself in English, which he never spoke with perfect fluency. But what he says contains so much that gives us an impression of the early days of their relationship that parts of it are worth reproducing. So in order to set the stage, let Bohr himself introduce to us some of the main characters while hinting at the plot.

Before the interview began, Bohr had talked informally to Kuhn and Rüdinger about his philosophical conceptions, and some of the questions addressed to him derive from this conversation. He opens the interview himself:

NB: Now we were just speaking about a kind of philosophical attitude one took at the earlier dates, and I tried to explain to Professor Kuhn that in some way I took a great interest in philosophy in the years after [high-school] student examination. I came especially in close connection with Høffding. That was just a minor thing, but I pointed out to him that there were some errors – actually there were many errors – in his formal logic. He took that to heart, and there came out a new edition, where he says that he has got some various help from one of his students....

AaP: Do you remember the kind of errors?

NB: No, but ... perhaps we shall find that edition of Høffding's, so we will see whether it says what kind of errors it was. They were really fundamental, not small things; but he was also not an expert in logic, so these things were just an incident.

The errors Bohr refers to here were those which Høffding had made in his book on formal logic, and Bohr seems to have called Høffding's attention to them whilst attending his propaedeutic course in philosophy at the University in the academic year 1903–1904. A couple of years later Høffding planned a new edition and, as we shall see in Chapter II, he contacted Bohr asking for his assistance. I would not claim that his errors were as elementary as Bohr seems to think. However, we learn from what Bohr here says that he already at that time regarded himself as one of Høffding's associates and that his interest in philosophy was aroused around that time, presumably by Høffding, though he does not state this explicitly. We also learn that Høffding obviously had a high regard for Bohr, who might otherwise have been thought rather impudent for criticizing his professor while just a beginning student.

His interest in philosophy was such that Bohr considered writing a book on philosophy himself. In the part of the interview that follows Bohr turns to this subject:

NB: At that time I really thought to write something about philosophy, and that was about this analogy with multi-valued functions. I felt that the various problems in psychology – which were called the big philosophical problems, of the free will and such things – that one could really reduce them when one considered how one really went about them, and that was done on the analogy to multi-valued functions.

Bohr then goes on to talk about various multi-valued functions of complex variables being mapped onto Riemannian sheets. In order to avoid ambiguity regarding which value of the multi-valued function is being considered, G.T.B. Riemann had proposed that for each value of the independent variable, the variables be severally mapped onto different planes by letting each plane represent a different set of values of a single valued function. After talking for a while about these functions, Bohr returns to the question of free will:

NB: Now, the point is, what's the analogy? The analogy is this, that you say that the idea of yourself is singular in our consciousness – (....) then you find – now it is really a formal way – that if you bring this idea in, then you leave a definite level of objectivity or subjectivity. For instance, when you have to do with the logarithm, then you can go around; you can change the function as much as you like; you can change it by 2π when you go one time round a singular point. But then you surely, in order to have it properly and be able to draw conclusions from it, will have to go all the way back again in order to be sure that the point is what you started on. – Now I'm saying it a little badly, but I will go on. – That is then the general scheme, and I felt so strongly that is was illuminating for the question of the free will, because if you go round, you speak about something else, unless you go really back again [the way you came]. That was the general scheme, you see. Will you ask something?

TSK: Yes. How did problems of this sort come to you in the first place? With whom did you talk about the problems like the free will?

NB: I don't know. It was in some way my life, you see. And I talked with somebody. But this was also extravagant, you see, so I think I did not learn something from other people, I just tried to show how close the analogy between our consciousness and such functions were, and that was really very close, indeed, but as just a help. What I was prepared for was that when we use any kind of word, this word has a certain connection with a certain degree of objectivity, and that you had to go back again, all the same way, in order to show what you could do with it.

AaP: Did you write anything down about this analogy?

NB: No, but I was very occupied with it. No, I did not write anything down, but I spoke to various people who came here. That was what I spoke with Kramers about, you see. Of course it was a kind of luxury, but it was also helping to [find out] what to do.

Bohr's thoughts seem here to be disconnected. Hendrik Anthony Kramers came to Denmark in 1916 and was Bohr's assistant for the following ten years. But, as we noted above, Bohr had already intended to write a book on this subject when still a student.

In his book on Bohr, Henry Folse has made a valiant attempt to reconstruct this extremely obscure analogy.[4] He believes that Bohr's analogy was intended to show that the problem of free will arises from an attempt to describe the psychological processes behind a human action in the same language as is used in describing the agent's experience, in that the experiencing subject cannot describe as an object its own experiencing activity. If we utter the sentence "I did the action A of my own free will", then the "I" to which we refer is what Kant called the "transcendental ego", which is always a subject and therefore cannot be considered as an object. Hence, neither can it enter into a causal account of the action. Instead we have two rival descriptions: that of the psychological account of the choice involved in a human action expressed in causal terms and that of an individual who reports an experience of freedom in spontaneously choosing to perform an act. But there is really no conflict here, Bohr seems to suggest, if we are attentive to the fact that the object of description is different in the two accounts owing to a shift in the level of objectivity in the context of which each of the descriptions is to be understood. Folse concludes quite correctly that Bohr was apparently not aware of how close his "solution" to the problem of free will came to that of Kant, for he was putting forward what he considered to be a totally original contribution.

I shall show that Bohr's interest in the problem of free will was something he had acquired from Høffding, who had written about this problem in many different contexts, and who had provided an analysis similar in certain respects to that of Kant on the incompatibility of the psychological and the ethical account of the action of a person. This assumption is partly confirmed a little further on in the interview when Bohr is asked when he started thinking about free will.

TSK: Did this first group of ideas about free will first come to you at the university before you started the work on surface tension?

NB: I think it was in those years before I got so [busy].... I was not really a kind of day-dreamer. I was prepared to do some very hard work, and this surface tension was a very great amount of work. Whether it's good or not, that's something else. But in between I was just interested also as regards the problems of biology, just what the problems of teleology meant, and so on. Therefore, I meant only that it was a natural thing to me to get into a problem where one really could not say anything from the classical point of view, but where it was clear that one had to make a very large change and that one got hold of something which one really believed in.

TSK: Did you carry on your interest in these problems by reading books of philosophy?

NB: No, not at all. (Laughter) Of course, I felt that philosophy – But that is my error, you see. It is not an error now but it was an error those days. I felt that philosophers were very odd people who really were lost, because they have not the instinct that it is important to learn something and that we must be prepared really to learn something of very great importance. And therefore in some ways I felt a long with what you were saying that the philosophers in Denmark –but I think they are the same, in principle, in Oxford and in the United States – There are all kinds of people, but I think it would be reasonable to say that no man who is a philosopher really understands what one means by the complementary description. I don't know if it is true, you see, because one can tell all kinds of people, and time goes, ... I think at any rate here [in Denmark] the thing is preposterous ... I do not also know how the things are there [in the United States]. But if you take it on a whole, a few years ago, they did not see that it was an objective description, and that it was the only possible objective descriptions.

Let us pause for a while. Bohr started his work on surface tension in the latter part of 1905, after the Royal Academy of Sciences and Letters had arranged a competition for papers on the experimental examination of the surface tension of liquids. He carried out his work in his father's laboratory and submitted his paper in 1906. The following year Bohr was awarded the gold medal by the Royal Academy for his entry, as was another young physicist, P.O. Petersen, who also submitted a paper. But, as Bohr says, his interest in free will was aroused before he started to work on surface tension. My suggestion, which I shall substantiate in Chapter II, is that Bohr's interest in these problems arose at the time when he was attending Høffding's introductory lectures on philosophy in 1903–1904, or possibly his lectures on the psychology of free will in the first part of 1905. I will show that this suggestion can be amply justified as far as what Bohr says here goes.

We also hear about Bohr's early interest in biology and the problem of teleology. Bohr's father, Christian Bohr, was a physiologist and professor at the University of Copenhagen, and was very interested in the dispute between vitalists and mechanicists in biology.

However, Høffding was also very interested in the problems of biology, about which he had already spoken at a meeting of the Biology Association in 1898. Indeed, he perceived teleology versus mechanism, and free will versus determinism, as two aspects of the same general complex problem of describing an individual whole (person, organism or living system). The talk he gave was published in 1905, around the time indicated here by Bohr, and Høffding could quite well have drawn Bohr's attention to it. Moreover, Høffding was a close friend of Bohr's father and, as we shall see, Bohr had the opportunity at an early age of becoming familiar with philosophical questions when his father and Høffding, together with other friends, gathered in the home of his childhood. It will be shown, in Chapter VI, that Bohr's conception of biology was very similar to that of Høffding and his father.

Bohr seems to have felt a certain animosity towards philosophers in general. Just after the above quotation he added, "It is hopeless to have any kind of understanding between scientists and philosophers directly". This feeling certainly springs from his lack of success in explaining complementarity to philosophers. He had had numerous fruitless discussions with Professor Jørgen Jørgensen (1894–1969), for many years the only Danish philosopher with an interest in the exact sciences and an important representative of logical empiricism. Jørgensen, one of Harald Høffding's younger students, was appointed professor of philosophy at the University of Copenhagen in 1924. For a period of nearly forty years he was the leading figure in Danish philosophy, making a great impact on more than one generation of scholars. His early interest in the neopositivism of the Vienna Circle and its phenomenalist foundation did not last, and little by little he moved towards a position which might be called critical realism. So, in spite of his vast knowledge of physics and mathematics Jørgensen did not think very highly of Bohr's philosophical ideas. And, since Jørgen Jørgensen must have been the one philosopher with whom Bohr had vehement discussions on complementarity over the years, it is little wonder that he sounds so despairing when philosophers are mentioned.

The late Jørgensen argued that quantum theory might be true with respect to what has been observed up to the present time, but from this we may not conclude that physicists will not one day be able to present a theory which describes atomic objects as they are in themselves independent of our observation. This apparent clash of opinion resulted in a long-running debate between Bohr and Jørgensen, and their discussions were in certain respects similar to those which Bohr had had earlier with Einstein.

There is an anecdote about one of the sessions between Bohr and Jørgensen on the interpretation of the new epistemological situation which had arisen in quantum mechanics. In commenting on what had been said Jørgensen is reported to have said to Bohr, "Professor Bohr, I have to admit that I understand nothing of what you are saying here", to which Bohr allegedly replied, "Well, Professor Jørgensen, I, on the contrary, understand everything you say". Whether the story is literally true or not doesn't really matter. It shows us that at least one thinker who had formerly been a prominent figure of the Vienna Circle was never able to grasp the essence of complementarity. Jørgensen, for one, apparently never saw any distinctively neopositivistic doctrines in Bohr's outlook, and certainly no realistic doctrines in it either. Moreover, if true, this anecdote indicates that Bohr never regarded himself as being in alignment with the school of logical positivism or thought that he held ideas quite similar to theirs.

Later in the interview Thomas Kuhn asked Bohr whether he had been introduced to the notion of a Riemann surface at school or at the University, and Bohr replied that he had learned about it at the University from a mathematician, Julius Petersen. Then Aage Petersen put the following question to Bohr, returning once more to the problem of free will:

AaP: Could I ask how the problem of free will was usually discussed then?

NB: I don't know, and I am very sorry what I have started on, but perhaps I will try to clear my thoughts another day. But the thing was that it was not a question. But ... everyone knew that it was a trouble, and that it did not fit in with classical physical ideas, and therefore, one wanted a broader scheme to put such questions in. I think it was also not too good, but I think it was an idea [I had] by myself which I really did not discuss, perhaps with my brother, but I felt just that it was a kind of escape or solution.

AaP: How did you look upon the history of philosophy?

NB: History of philosophy?

AaP: What kind of contributions did you think people like Spinoza, Hume and Kant had made?

NB: That is difficult to answer, but I felt that these various questions were treated in an irrelevant manner.

AaP: Also Berkeley?

NB: No, I knew what views Berkeley had I had seen a little in Høffding's writings, and I thought it was obvious that so could one do it, but it was not what one wanted.

TSK: Did you read the works of any of these philosophers themselves?

NB: I read some, but that was an interest by – oh, the whole thing is coming [back to me]! I was a close friend of Rubin, and, therefore, I read actually the work of William James. William James is really wonderful in the way that he makes it clear – I think I read the book, or a paragraph, called – No, what is that called? – It is called "The Stream of Thoughts", where he in a most clear manner shows that it is quite impossible to analyze things in terms of – I don't know what one calls them, not atoms. I mean simply, if you have some things ... they are so connected that if you try to separate them from each other, it just has nothing to do with the actual situation. I think that we shall really go into these things, and I know something about William James. That is coming first up now. And that was because I spoke to people about other things, and then Rubin advised me to read something of William James, and I thought he was most wonderful.

TSK: When was this that you read William James?

NB: That may be a little later on, that would be ... I don't know. When I got into ... When I got so much to do, and it may be at the time I was working with surface tension, or it may be just a little later. I don't know.

TSK: But it would be before Manchester? I mean it was still ...

NB: Oh yes, ...

TSK: ... as a student

NB: ... Oh yes before, it was many many years .... Not many years, but I mean ... [background noise makes the tape inaudible]. You see, the problem is so difficult, and it may be even irrelevant and immodest to speak so, but I was not interested in philosophy as one generally called it, but I was interested in this special scheme, and that was even not too good.

This passage is much cited as evidence of William James's influence on Bohr. But, as we shall see in Chapter II, Bohr's memory on this point has been called into question by Léon Rosenfeld, who had the distinct impression that Bohr first read James around 1932. However, in spite of his age Bohr's own testimony must be given greater weight unless additional evidence to the contrary should emerge. But no such independent evidence seems to exist. Instead, I shall argue that certain facts do in fact lend support to Bohr's testimony; but this does not mean that I think he was deeply influenced by James's philosophy. Rather, I believe he was taught, by Høffding, ideas similar to those of James. As will be demonstrated, Edgar Rubin, who is thought to have encouraged Bohr to read James, had been a friend of his since their undergraduate days; and it was he who set up, around the time Bohr here indicates, a student club, *Ekliptika*, of which they both were members, whose aim it was to provide a forum for the discussion of Høffding's lectures.

Several writers have likewise suggested that Bohr might have been inspired by Kierkegaard. The idea that Bohr was influenced directly by the Danish philosopher Søren Kierkegaard seems, as I shall argue in Chapter II, not very likely and very difficult to substantiate; but Kierkegaard may have had an indirect impact on Bohr as a young man through Høffding's books and lectures as well as through some of Høffding's views on psychology and the theory of knowledge which stem from Kierkegaard.

After talking about William James, Bohr spoke of Høffding once more in answer to a question put by Kuhn. Here he referred to one of the many visits he had paid Høffding at the Carlsberg Mansion, which was his own residence at the time of the interview. The episode he mentions took place at a time when Høffding was ill.

TSK: Did you often see Høffding?

NB: Oh yes, I had very much to do with Høffding. He had some difficulties, and I came out here and tried to read him about poetry. That was Wildenvey, a Norwegian writer. And Høffding was really very interested [in complementarity], far more interested than any philosopher who has been called a philosopher, because he thought it was right. He had not too great an understanding of it, but he wrote an article about these things, which is far better than any other thing which has appeared in philosophy since. Perhaps it is wrong.

AaP: Oh, I think that is to go too far.

NB: No, I think that it is not too far; it may be that it is not good.

AaP: Well, he wrote mainly about his own anticipations of these ideas. ...

NB: It is an odd thing. First of all, it is not at all meant to be an objective description, and philosophers may be much, much better than I think they are, but actually, now it is thirty-five years since one really got the .... [answer]. But I speak about the time, let us say up till after the war [since when] there may be some better, but I do not know what their names are. Then the philosophers simply were critical, but Høffding was not critical. — I don't know. That is a very difficult thing, and that is also a thing we shall not go into, but it would be nice to ask you, how it really is with the philosophers.

Once again we get a forcible impression of how disappointed Bohr had been by philosophers and their resistance to his ideas.

The paper by Høffding which Bohr mentions was published in 1930, and was Høffding's last work. In it he gave, as we shall see in Chapter III, an account of the recent developments in quantum mechanics and stated that the idea of complementarity contains nothing but what he himself had described in psychology. This is what Aage Petersen comments on. When Bohr says "Perhaps it is wrong", it is not easy to see what he means. Hence Petersen's interjection also becomes ambiguous. The word "it" may either refer to his idea of complementarity or to Høffding's paper. It seems most reasonable from the context, I think, to construe the remark as referring to Høffding's paper. So what Bohr intends to say is that even though he thinks Høffding's paper satisfactorily expresses the author's positive attitude to the ideas of complementarity, the presentation was not cogent enough to convince other philosophers. Thus Petersen's response seems to suggest that he thinks it is not correct to say that Høffding's article was "better than any other thing which has appeared in philosophy since", while Bohr takes him to mean that it is wrong to say that the paper might be "wrong".

The remainder of the interview (the whole transcript covers 10 pages) deals with the discussions had by Bohr and Einstein. Einstein had, from the first, taken a critical stand with regard to quantum physics, thereby opposing Bohr's interpretation of quantum mechanics. He strongly felt that orthodox quantum mechanics would prove to be a premature theory. First he attempted without success to contravene quantum theory itself by appealing to an inconsistency between the predictions of quantum mechanics and the outcome of various thought-experiments, but in discussions with Bohr the latter was able to show that in each case the outcome would in fact be in agreement with quantum mechanics. Later Einstein held that quantum theory, though consistent, yielded an incomplete description of nature and he produced, together with Podolsky and Rosen, an argument based on a certain thought-experiment in order to vindicate his own framework of realism. But in Bohr's view such a framework was out-dated: Einstein "simply took the view of old-fashioned philosophy, took the view of Kant", he says.

With respect to the earlier debate with Einstein, Bohr mentions that he immediately saw the way to answering to Einstein's objections. What he has in mind here is the following. At the Solvey meeting in 1930 he turned General Relativity into a deadly weapon against the criticism made by Einstein by proving that the latter had not paid attention to what effects would be generated according to his own theory, and if he had, he would have discovered that the predicted effects were in agreement with Heisenberg's uncertainty relations. But with respect to the EPR-paper Bohr is not quite right when he asserts "one had also to think [just?] a little to see what the solution was". As I attempt to prove in Chapter VII, no single paper ever had such a great impact on Bohr as did this. It induced Bohr to change his philosophy of complementarity from being one originally arising from reflections on the limitations of the measurability of observables to one which is grounded on reflection on the logical requirements for the definability of these observables.

It is perhaps ironic that Bohr here makes a comparison between Einstein and Kant in relation to their view of nature. For many commentators on Bohr such as C. F. von Weizsäcker, C. A. Hooker, Henry Folse, John Honner and Dugald Murdoch have pointed to similarities between Bohr's own thought and that of Kant. But both readings of the facts of the case can be shown to be correct. I suggest that the following Kantian elements of Bohr's philosophy all have their source in Høffding's anti-realist philosophy: the indispensability claims he made for classical concepts, his conception of the subject-object distinction, his criteria for what it is for something to be real, and his notion of phenomenon. Likewise I shall argue that the elements of Kant's theory which Bohr dislikes, and which he believes can be associated with Einstein's view, are also those which Høffding repudiates: namely, those according to which it makes sense to talk about a realm of transcendental objects behind the phenomena. Bohr, apparently, saw Einstein's efforts to interpret the state vector as an expression concerning only a statistical ensemble of objects and not the physical state of a single object, as an attempt to introduce the idea of objects as possessing properties which are inaccessible to human experience, but which would make the description of them deterministic. Such an interpretation, however, did not fit in with Bohr's understanding of an objective description as a description that refers only to what can be related to our experience.

At the end of the interview Bohr appears to be tired, and the session closes with a question from Kuhn, who once again returns to Høffding.

TSK: Let me take you, if I may, back to the very beginning again. Would you tell us just a bit more about your early relation with Høffding? Just what sort of a person was he?

NB: He was a very fine person. First of all, he was an imposing person in the way of understanding, and he was the best "*kender*" [expert], I think, of his time, of Spinoza and such things. – I think I must stop, but I can tell you a little bit of a story about

> him. He had so many sorrows, when he was old. He married when he was more than eighty, for the second time, and his wife really died in an asylum before he. I went out an evening in all these troubles, when his wife was in an asylum, to try to cheer him up, and brought out some poems of the Norwegian poet Wildenvey, a very philosophical kind of poet, to read for him, which I also did. Then we were sitting in our dining room, and having some tea, and then there is a statue in the room of Hebe, which carries the nectar of the gods. He suddenly said to me, "If I had realized how difficult it is really to get to know what the sentiment of Hebe is, whether she is mild or severe". – You see, that depends, just on what one likes to do in that kind of statue. – But he added, that he lived upstairs, and every morning when he came down, he looked up to Hebe to see whether she was satisfied with him or not. That seems a very odd story; it is a very beautiful story, because he took things very seriously.

Here the interview ends. It is striking that throughout the conversation Bohr again and again mentions Høffding and Einstein by name but only mentions other people once or twice. A psychological explanation is both called for and welcome. No persons other than Høffding and Einstein have meant so much to Bohr from an intellectual point of view. For whole periods of his life he had lived day and night with these two men, expending a lot of mental energy and intellectual resources on understanding their theories and arguments. Einstein as the first among peers, whose approval Bohr had wanted more than anything else but, as is well-known, never gained; Høffding as his mentor and the person who was to initiate Bohr into philosophy and whose respect and acceptance he easily gained, as will emerge in the course of this book.

But, one might ask, of what interest is it to know that Bohr was influenced by Høffding? I argue it is of more than purely biographical interest, though such interest would in itself be entirely legitimate. Since Bohr's ideas on complementarity have played a dominant role in the understanding of atomic phenomena from the dawn of quantum mechanics, it is also, from the point of view of the history of science, of interest to trace their roots back as far as possible. But this is not all. Bohr never presented his ideas of complementarity systematically; rather, they were developed in connection with talks he gave all over the world. He therefore analyzed neither their epistemological nor their ontological implications to any great extent. Furthermore, Bohr's style is so rugged and uneven that in many places his writings are not free from obscurity. For although Bohr naturally tried very hard to express himself with precision, as his collaborators and admirers have emphasized, his style is, nevertheless, opaque and taxing. It is correct that some of problems of understanding Bohr's texts are due to the fact that over the course of fifty years his writings display a gradual development and refinement which result to some extent in terminological variations and inconsistencies. But, apart from those, there are also certain signs in his manuscripts which indicate that he felt a personal inadequacy in regard to expressing his thoughts in writing, and his difficulties in this regard were compounded by the formidable task of formulating an interpretation of quantum mechanics coherently. But, what is more important, a fact that has not been much noticed, is that during the thirties after his debate with Einstein, Bohr modified some of the underlying arguments for his philosophical outlook. These factors combined have led to the emergence of a plurality of interpretations as well as misinterpretations of Bohr's philosophy.

However, I think that a reliable means towards obtaining a grasp of the notion of complementarity that eliminates misunderstandings is to expose the origin of Bohr's philosophy in order to display its legacy. Indeed, I will go so far as to say that any interpretation of Bohr's philosophy is made more intelligible when understood in the light of an account of the philosophical milieu in which he grew up and of the impact it had on him. Hence, our task will be first to uncover what Bohr inherited from his predecessors in order to elucidate our understanding of his philosophy and then to consider what we today, in turn, have inherited from Bohr. The point is to try to understand Bohr better, and that we can do much more effectively, I think, when we see just what Bohr was taught in the manner of a philosophical vocabulary and of problems as Høffding perceived them. However, I also argue that the philosophical vindication which Bohr gave his interpretation underwent a modification after Høffding's death due to the challenge of the EPR thought-experiment. Bohr's reaction to this argument resulted in what I suggest is in fact the legacy of Niels Bohr.

Ultimately, the question is, of course, whether an attempted demonstration of the influence of Høffding on Bohr is going to be successful: to what extent and with what arguments can a claim of a philosophical debt be established? It is quite obvious that when an individual acknowledges either in public, in speeches or interviews, or in private, in letters or diaries, that he owes some or many of his ideas to another, historians have the strongest possible evidence they can ever have to support the claim that there exists such a debt. On the other hand, if an analysis of two philosophers' ideas does not show any or only very little similarity, historians will probably dismiss the evidence regardless of what the individual himself tells us. So even a person's own testimony will not be considered as proof but merely as evidence, although of the strongest possible kind, of the debt. It is very rare, however, that historians will find personal statements which explicitly acknowledge intellectual debts. But this does not prevent them from arguing that such influence was present. The fact that Plato in his youth was heavily influenced by Socrates is not something we know because Plato has told us so. We know it from the reading of Plato's works and those of other writers of that time, and from learning that Socrates was Plato's teacher and an intimate friend of his.

Thus, two conditions are necessary as well as sufficient for an intellectual influence rightly to be claimed to exist. There must be an epistemic correspondence between two persons' ideas and, further, there has to be a causal connection between these ideas. Records of frequent encounters, of a friendship, of teaching, of the one reading or hearing about the other's ideas, and the existence of a similarity between their ideas will thus all be evidence to support a hypothesis of the presence of an intellectual influence exerted by one on the other. But whether or not the evidence is sufficient to confirm the presumption is a matter that rests with judgment. It is impossible to find criteria which specify once and for all how much evidence of this kind is needed to back up a claim. Ultimately, it will depend on a personal assessment as to whether or not the evidence shows that the ideas are sufficiently similar and that the causal connection is sufficiently well-documented for us to conclude that a person has acquired some of his ideas from another individual.

With respect to Bohr and Høffding I shall argue that the evidence is strong enough to satisfy both conditions. We shall see (i) that there were encounters and intellectual exchanges between them for over three decades; (ii) that Høffding taught Bohr philosophy; (iii) that they were close friends; (iv) that they discussed philosophical matters regularly over the years. But more important than the mere fact that Bohr participated in discussions with Høffding is that it seems obvious that his knowledge of philosophy, its history and its problems – and in particular its relation to psychology – all of this was filtered through the vocabulary and problems and interests that characterized Høffding's philosophy. So, in addition to the evidence of their frequent discussions I shall point out the similarities between Høffding's and the young Bohr's ideas on epistemology, psychology and biology. This evidence should be considered together with the fact that several of Bohr's former assistants testify that Bohr found it very gratifying that his interpretation of quantum mechanics harmonized with his earlier ideas on psychology and epistemology.[5] Thu, on an evaluation of all this evidence, it is very reasonable to believe that Bohr must have acquired from Høffding a quite specific pattern of thought which led him to the idea of complementarity and shaped the formulation he gave his new view.

PART I

Høffding as Mentor

Chapter I

1. HARALD HØFFDING, HIS LIFE AND THOUGHT

Harald Høffding saw the light for the first time on 11th March 1843, and eighty-eight years later, on 2nd July 1931, for the last. He was the third son of a wealthy and respectable founder of a trading house, N.F. Høffding, in the Danish capital, Copenhagen. When Høffding was born, Copenhagen had a population of only 200,000 inhabitants. Denmark had then just become a smaller country through the loss of Norway to Sweden in 1814 and was to become even smaller through its defeat by Germany in 1864 and the resultant cession of Slesvig-Holstein to the Germans.

Høffding lived in his parents' home until in 1870 at the age of 27 he married his first wife, Emmarenzia Lucie Pape. Unfortunately she lived for only seven years subsequent to their marriage, leaving her husband with two sons. For a period he lived with his sister-in-law, who had lost her husband in the same year as Høffding had been widowed. Høffding remarried in 1924, his second wife being a young Swedish admirer, Greta Sofia Maria Ellstam, who was at that time only 24 years old, while he was eighty-one. It was not a happy marriage and brought Høffding much sorrow. His wife was unbalanced, and ultimately this led to a severe mental disorder. She was put into an asylum where she died in 1930, a year before her husband.

As a child Høffding was a leader among his class-mates. He went to Mariboe school for the first four years of his schooling, and then he attended the nearby "*gymnasium*" "Metropolitanskole" in the middle of the city. He passed the "*studentereksamen*" (the equivalent of a high-school diploma) with distinction in 1861. Lessons in Greek coupled with his interest in Plato's philosophy gave rise to a lifelong love of Greek culture. In his final years at the "*gymnasium*" he attended some lectures at the University of Copenhagen: those in aesthetics given by Carsten Hauch (1790–1872), an eminent Danish poet and professor of aesthetics; and those in theology given by B.J. Fogs (1819–1896). Although the University of Copenhagen, which was the only one in Denmark at that time, had a history going back to 1479 it was still a very small university with about 40 lecturers in all. When Høffding was enrolled at the University in 1861 he chose theology for his subject. This choice was motivated not by parental influence but by a deep interest in religious questions deriving in part from the preparation for his confirmation and in part from his compelling urge to confront and examine existential problems.

Parallel with his theological studies Høffding had to take a compulsory one-year course in propaedeutic philosophy ("*Filosofikum*"), which at that time comprised ten hours of lectures per week. He attended the lectures of Frederik Christian Sibbern (1785–1872) on logic and psychology and those of Rasmus Nielsen (1809–1883) on a general introduction to philosophy. Both were professors of philosophy. And after taking the examination in philosophy in 1862 Høffding sustained his interest in the subject by attending lectures given by Rasmus Nielsen and Hans Brøchner (1820–1875), who was his third teacher of philosophy and who succeeded Sibbern as professor in 1870.

The philosophy of Sibbern was formed in the main by his attitude toward and criticism of Hegel's philosophy.[1] Sibbern aimed at developing a metaphysics which was an alternative to that of Hegel both with respect to his philosophy of nature and his psychology. He rejected Hegel's idea that existence is identical with the absolute spirit. He contended that within a semi-materialistic ontology each individual spirit possesses the capacity to acquire more or less correct knowledge by "organizing" itself into the whole, a process subject to the law of development, which had it that every cognizing individual continually has to take its provisional assumptions up to review. In psychology Sibbern's approach was that of seeing the mind as a totality. Classifying mental capacities under the heads of cognition, emotion and will, he believed that mind and body were separate effects of one and the same cause. The tripartition of the mind and the view of mind and body as two aspects of the same substance or process, at that time called the identity hypothesis, were theories Sibbern had passed onto him by his teacher Niels Treschow (1751–1833), who had acquired these ideas from Kant and Spinoza, respectively.[2] Treschow had regarded the will as the most fundamental and original of the three. He had also posited a principle of personality which stated, contrary to the ideas of British empiricists, that the mind consists of a nucleus which figures in the laws of association. Each person has to be regarded as a unity, according to Treschow, a view we encounter in Høffding's thought.

Høffding writes in his *Memoirs* that he had looked forward to following Sibbern's lectures but that, since Sibbern was advanced in years, the lectures proved very disappointing. Later, when Høffding himself had become a philosopher, he came to appreciate Sibbern through seeing his own psychological studies as a continuation of his former teacher's, especially with regard to the emotions, although he accused Sibbern of anchoring his psychology to a speculative foundation.

Rasmus Nielsen had started out as a pupil of Hegel, whose philosophy had become fashionable in Denmark during the 1830s through introductions from the pen of J.L. Heiberg (1791–1860), a well-known author, and of H.L. Martensen (1808–84), a professor of theology and later bishop. The latter was attacked vehemently by none other than Søren Kierkegaard (1813–1855). Like Martensen, Rasmus Nielsen tried to establish Christian Dogmatics on the basis of the philosophy of Hegel. But he then fell under the influence of Kierkegaard, first through his works and later through conversations with him. As a result of this influence he reached the conviction that Christianity could not be a subject for speculative understanding. Faith and knowledge belonged to two separate levels or planes in such a way that mutual contradiction could not occur. But Rasmus Nielsen attempted to prove this from speculative premises, and this brought about a rupture in his relations with Kierkegaard.

As a student Høffding underwent a religious crisis; he also devoted himself to the works of Kierkegaard as Rasmus Nielsen had done before him. After a hard struggle Høffding felt forced to agree with Kierkegaard that the demands on the individual made by Christianity in its original simplicity cannot be met in a life molded by family and state, art and science, and that the Church had betrayed Christianity. He became ultimately convinced that Kierkegaard had shown, and with biting irony, that the compatibility of science and faith, which he had previously believed Martensen to have accomplished by unifying Hegel's philosophy and Christian Dogmatics, was an illusion. During this period he oscillated between theology and philosophy. He ultimately found himself faced by the requirement that Kierkegaard claimed to be incumbent on every candidate for the ministry and felt unable to meet it: he admitted that he couldn't live up to the ethics of primitive Christianity. And he decided not to enter the ministry. He nevertheless continued his theological studies until he graduated in theology in 1865.

Kierkegaard had a lasting influence on Høffding. His dictum that "Subjectivity is the truth" is one aspect of it. Another is his principle of personality. Høffding writes in his *Memoirs*, "The study of Kierkegaard introduced me to an idea which subsequent philosophical studies led me to amplify and give particular application. This was the notion that the formal feature of the life of the mind is to be found in the unitary and convergent aspects of its synthesizing of experience. And that the measure of an intellectual life resides in the relation between the compass of its content and the dynamism with which this is brought into focus".[3] A similar principle of personality was part of the psychological tradition deriving from Treschow and Sibbern, whose student Kierkegaard had been.

Høffding calls the years of crisis in his *Memoirs* "the most difficult and darkest in my life". He lived a life of austerity and was very often despondent and withdrawn, giving his family cause for anxiety. But the decision not to take holy orders and his diligence in working for his examination had a stabilizing effect; and just after passing the examination in 1865 he made a decision which was to be the most important turning point in his life: he became engaged to Emma Pape.

Instead of entering the ministry then, he turned to teaching and taught at his former school, where he remained for the following 17 years. However, most of his spare time was spent on philosophical studies. He responded in 1868 to a prize paper set by the University entitled "*Hvorvidt kan den i vor Litteratur i sin Tid førte Strid om den frie menneskelige Villies Realitet siges at have ført til et blivende og udtømmende videnskabeligt Resultat*" (To what extent the former dispute in our literature about the reality of the will of free human beings can be said to have led to a conclusive and final scientific result). Here we see Høffding for the first time faced with the problem of free will to which he was to give great attention during the rest of his life. His paper was deemed to merit the gold medal; but much later, in his *Memoirs*, Høffding was to declare that he was not wholly satisfied with the solution he had adumbrated. The value of his paper lay, according to the examiners, Rasmus Nielsen and Hans Brøchner, in the thoroughness of his characterization of the various positions which were taken in the so-called Howitz-controversy in the 1820s with respect to determinism versus indeterminism. Høffding continued his work on the concept of free will, however, in connection with his studies of Greek philosophy, and two years later in 1870 he obtained his philosophical doctorate with the thesis *Den antikke Opfattelse af Menneskets Villie* (The Conception of the Will in Antiquity).

In 1866, only a year after his graduation, Høffding published his first paper, a contribution to the debate concerning faith and knowledge which had begun two years earlier when his former teacher, Rasmus Nielsen, published the first part of his book on the logic of basic ideas. Høffding here defended Rasmus Nielsen's view that science and the Christian faith were so essentially different that they did not contradict one another, but each represented a kind of truth. But one year later Høffding wrote a letter to his fiancee in which he declared that he now opposed Nielsen's dualism partly because of the impact of the criticism of a former teacher, Hans Brøchner. In the same letter he added that he was quite certain that Christianity would always represent for him the highest truth. This conviction did not last long, however. He subsequently abandoned his Christian faith, although retaining a deep interest in religious questions for the rest of his life.

During the following years Brøchner had a growing influence on Høffding and they were close associates. After Brøchner's death in 1875 Høffding published, inter alia, Brøchner's Memoirs of Kierkegaard. Brøchner's philosophical roots were essentially Hegel's speculative method and thought. But what may have made an impression on Høffding were Brøchner's attempts at developing a view of life which, more earnestly now, focused on "the real", the natural man, and on knowledge of science. Brøchner also worked on the history of philosophy, and in 1856 he published a monograph on Spinoza, of whom Høffding in his *Memoirs* says, "If I am to call myself after somebody, I will call myself after him", reiterating Lessing's words with respect to Spinoza. So it will come as no surprise to learn that many years later Høffding too wrote a monograph on Spinoza.

In 1868 Brøchner advised Høffding to go abroad. This suggestion was later to become a significant one for Danish philosophy. Høffding hesitated between electing to go to Germany, whose science and philosophy had had such a grip on Danish culture for about a century, or to go to France. Fortunately, he chose to stay in Paris during the winter of 1868–1869. As he puts it in his *Memoirs*, "Here I learned about other schools of thought than those emerging from Germany, which until now have had the greatest influence on those who practiced philosophy at home".[4] He attended lectures by Taine and read Comte and Spencer; thus, he became aware of the tenets of positivism which gradually became part of his outlook. Later he introduced positivism to the reading public at home, at first through the book *Den engelske Philosophi i vor Tid* (English Philosophy of Our Time) from 1874 as well as through his lectures on Charles Darwin and Herbert Spencer, two of whose works he translated around that time.

Not until 1880, when he was appointed reader, did Høffding hold his first university chair, and three years later he was elected as Rasmus Nielsen's successor as professor of philosophy. But in 1870 he claimed his right as a doctor of lecturing at the University on a private basis (*jus docendi*), a right of which he availed himself from the spring of 1871. By the middle of the decade Høffding had reached a position which he characterized both as critical positivism and as critical monism. This position gave at once satisfaction to his personal taste for clarity and supplied him with a philosophy which could fill in the gaps between science, art and religion. In a lecture in 1874 he gave an introduction to philosophical problems as he saw them at that time and to their interconnections. This lecture became a sort of philosophical manifesto for Høffding's future work.

It was essential for Høffding that his "Introduction to Philosophy" be based on philosophical problems. To him it was these and not principles or systems which constitute philosophy. He distinguished between four fundamental issues: 1. the problem of mind (the psychological problem), 2. the problem of knowledge (the logical problem), 3. the problem of existence (the cosmological or metaphysical problem), and 4. the problem of evaluation (the ethical-religious problem). The division is a result of Høffding's historical studies, inasmuch as he believed that this was the optimal way of integrating the problems which had exercised the minds of philosophers throughout the ages. However he later ranked the problems in a different order. Høffding believed, nevertheless, that they represent the same aspect of one and the same basic problem, which he identified as the relation between unity and plurality, connection and singularity, or between continuity and discontinuity as was his preferred way of putting it. Again and again he stressed that common to all problems of knowledge, of existence, and of evaluation is the incompatibility or the antinomy between continuity and discontinuity. This insight was drawn from two quite different sources. One was Kierkegaard's principle of personality, according to which that which characterizes the mind is a unification of experience and emotion. The other was the tradition of Comte and Spencer. Thus, in one context he refers to Spencer's formula of development as one implying that the nature of mental processes is that of aiming at synthesis, the nature of cognition is a striving towards conceptual coherence of a maximal number of phenomena with respect to a certain principle or theory, and that the goal of ethics is the creation of a rich and stable personality in harmony with the demands of the common good.

As mentioned above, Høffding's lecture of 1874 took the form of a philosophical manifesto whose motifs were to characterize his work throughout the rest of his life. He devoted his labors to all four problems, dividing his activities into three different periods. From 1875 to 1887 he worked on psychology and ethics, from 1887 to 1895 on the history of philosophy, and after 1895 on the philosophy of religion and epistemology. Psychology was the first fundamental problem to which he applied himself from 1875 to 1882, when he published a survey of psychology entitled *Psykologi i Omrids på Grundlag af Erfaringen* (An Outline of Psychology on the Basis of Experience). It was translated in the course of the following years into German, French and English. As the title suggests emphasis, in harmony with his new positivistic attitude, is laid on an analysis of the way in which we experience our own mind and the minds of others rather than on a metaphysical account. He denies the tenability of a substantival account of mind, arguing for a Spinozistic identity hypothesis as his predecessors had done before him. But even though Høffding bases his exposition on experience he is very critical of the use of experimental methods in psychology, holding that the experimental set-up intrudes too much upon the subject of investigation. This does not mean, however, that he did not draw upon experimental results in his own account; indeed he continued to add new relevant results to the later editions of the book. Høffding also regards associationist psychology with skepticism, opposing, for instance, any attempt to reduce associations of similarity to association by contiguity. Like Sibbern before him, he sees the mind as a synthesizing dynamic process, a synthesis – a name he borrowed from Kant – whose content could be very varied and multiple but nevertheless bore the mark of unity and continuity. Synthesis is the informing principle of the mind, according to Høffding.

In maintaining this theory Høffding in fact anticipated Gestalt psychology. In his "law of relation" he states that every element and every state of the mind is determined by the connections into which it enters together with other elements or states of the mind. It is impossible to analyze the mind into permanent elements because there are no such elements. What may be looked upon as an element according to one description will perhaps be seen as a compound under another description. The elements of the mind do not exist in isolation and can be separated only through abstraction.

Høffding accepts the trisection of the mind into cognition, emotion and will, which goes back to Sibbern, Treschow and Kant, claiming that will, in the broadest sense, is the most fundamental of these three. However, his account of, for instance, the passage from the unconscious to the conscious, memory, comparison and decision rest on analyses of his own. Moreover, Høffding drew attention to, prior to anybody else, the existence of an immediate quality of familiarity.

In the same year as that in which he completed the book on psychology, he spoke in "*Studentersamfundet*", the newly founded society of students, giving a talk entitled "Om Realisme i Videnskab og Tro" (On Realism in Science and Faith), which was later published in the collection *Mindre Arbejder* (Minor Works), part I. Here he appears as an adherent of "realism", which he defines as the principle of natural causes. Such a realism always deduces certain phenomena of nature from other phenomena of nature, the more complex phenomena from the more simple. Realism in this sense claims "that true knowledge does not consist of accumulated experiences but is insight into the interrelationships between experiences". This assumption brings realism closer to idealism, according to Høffding, since one of the essential ideas behind idealism is that we must not accept phenomena as single, isolated facts but that we have to find a bond which brings connectedness and unity. What is wrong with idealism, however, is that its adherents believe that this connectedness and unity can be found by following another path than that of mere continuous manipulation of the facts of experience. It is not possible, as idealism suggests, to gain access to what underlies reality through pure speculation. But via experience, as realism asserts, it is. Høffding also claims that such a realism is consistent with the efforts to maintain the value of intellectual life and its importance for reality, an idea essential to idealism. The realism of faith can thus, according to him, be defined as the position which holds the idea of an ideal value of progress together with the idea that this ideal value is realized through the effect of the natural causes.

Høffding's second main work was a book entitled *Etik. En Fremstilling af de etiske Principper og deres Anvendelse paa de vigtigste Livsforhold* (Ethics. An exposition of the ethical principles and their application to the chief circumstances of life), which was published in 1887 and later translated into French, German and Russian. After Høffding had abandoned Christianity he set about the task of constructing an ethics without a religious foundation. The basis of ethics had, indeed, to be found in natural causes in accordance with the programme of realism he had proclaimed in 1882; that is, the principles of ethics are to be based on human nature as made manifest by psychology. It is obvious that we have here to face the problem that, in actual fact, evaluations of men's actions differ with the result that these evaluations will thus in the end be seen to be subjective and can never have universal validity. Høffding accepts too that there exists no ethics which can be rationally imposed on everyone. There are various possible positions from which we can make an ethical judgment, from the perspective of the individual to that of mankind. But Høffding still believed that if we reflect upon history we will see that within the Greco-Roman culture to which we belong, universal sympathy and the common good have been leading motives behind man's actions, motives which have been elaborated through the influence of Christian ideas and through the humanistic view of man that arose in the eighteenth and nineteenth centuries. These values were integral to Høffding's ethical point of view: universal sympathy formed in his system the subjective principle and the common good formed the objective principle.

Høffding's humanistic attitude came very much to the fore in a controversy he engaged in with the famous critic Georg Brandes (1842–1927) in 1889–1890. In a paper about Nietzsche, Brandes had defended his "radical aristocratism", according to which the emergence of great personalities is the goal of history and the pursuit of the common good contemptible. Towards this Høffding reacted with a special plea for "democratic radicalism". Like Kant he could not accept that the majority should be a mere means in the service of the very few. Instead he adhered to the principle of personality to the effect that no human being ought to be regarded or treated merely as a means but always as an end in himself, a principle which he, unlike Kant and reflecting his empiricism, deduced from the principle of the greatest happiness of the greatest number.

A few years earlier Høffding had become politically active, which was tantamount to jeopardizing his position as professor. At that time the Danish government was, *de facto* and not merely *de jure*, nominated by the King, its members drawn mainly from the land-owning class, with the consequence that the government under the leadership of the Conservative prime minister J.B.S. Estrup (1825–1913) found itself in opposition to the parliament where the liberals had the majority. Estrup then ruled by provisional laws. In this atmosphere of political tension both Høffding and another younger professor, K. Erslev (1852–1930), a historian, addressed the "*Studenterforeningen*", the student union, and expressed their support in favor of some liberal politicians who had removed a chief constable from a platform which he had mounted in order to report about the meeting to the government. After a heated debate and a demand for their dismissal in the Conservative newspapers, the episode ended with a severe reprimand.

During the years 1887–1895 Høffding devoted himself to the study of the history of philosophy. His main work from that period is *Den nyere Filosofis Historie. En fremstilling af Filosofiens Historie fra Renaissancens Slutning til vore Dage* (A history of modern philosophy. A sketch of the history of philosophy from the close of the Renaissance to our own day), which was published in two volumes in 1894–1895, and as early as 1900 it was translated into English from the German edition and published in London. It testifies to both the breadth of Høffding's reading and his personal assimilation of what he had read. In particular, he lays stress on the philosopher's own cultural background as well as on the connection between philosophy and social and cultural life. Therefore not only philosophers but scientists such as Copernicus, Galileo, Newton, Darwin and Robert Mayer are treated in the book because they "have through their research initiated important changes in understanding of philosophical problems". Apart from this historical survey, which was expanded in another book, *Moderne Filosofer* (Modern philosophers), in 1904 and which was published in English in New York in 1913, Høffding also wrote various monographs on individual philosophers whom he admired, especially Spinoza (1877, 1918, 1921), Kant (1893, 1894, 1896) and Kierkegaard (1892, 1913).

From 1895 onwards Høffding devoted himself to the philosophy of religion and epistemology. The starting point of his philosophy of religion is the claim that religion has, since the rise of science, been unable to fulfil the human need for true knowledge. Religious experience does not arise from an intellectual basis, nor is its value of an intellectual nature. The task of the philosophy of religion is then to undertake a psychological investigation of the sources of religious ideas and feelings which, according to Høffding, have their roots in the relation between reality and value. What is essential to every religion is a belief that despite all changes in external reality, it is possible to maintain the values of life. However, religion can neither solve the enigmas of existence nor justify ethical norms. In questioning dogmatics, the best and most valuable part of religion may still be preserved as a form of undogmatic poetry of life expressing the loftiest thoughts and deepest feelings in terms of symbols and images. His most important book on religion, *Religionsfilosofi* (The philosophy of religion), was published in 1901. A few years later it was translated into English from German and published both in London and New York in 1906.

In the autumn of 1902 Høffding was elected vice-chancellor (or president) for the following year at the University of Copenhagen. In the same year he published his first major work on epistemology, called *Filosofiske Problemer* (The problems of philosophy), which was translated into English and printed in New York in 1905 and contained a preface written by William James. Høffding corresponded with James over a period of years, and he had stayed at his home twice the previous year, when he travelled to America and England. On the second visit Pierre Janet and Lloyd Morgan were also staying with James. Høffding tells us in his *Memoirs* that, shortly after he had left America for England, James had said to a correspondent from Oxford "make much of good old Høffding", he is "a good pluralist and irrationalist". However, Høffding rightly points out that there was nothing in the lecture he gave at Harvard which could humbug James into the belief that he was a pluralist but, as he himself states in the lecture which was published later, he sympathizes with pluralism with respect to its methodology and to a certain degree its metaphysics, though he would, nevertheless, call himself a critical monist because continuity and connectedness present themselves to him as being weightier.[5] After the death of James in 1910, Mrs. James wrote to Høffding, "William had a strong affection for you".

The book on philosophical problems was merely a prolegomenon to Høffding's fifth and last major work, *Den menneskelige Tanke, dens Former and dens Opgaver* (Human Thought, Its Forms and Its Tasks), published in 1910. The work brings together all the threads of his philosophy. The first chapter deals with the psychology of human thought. The following chapters deal with the forms thoughts take, the categories, which Høffding divides into the fundamental, the formal, the real and the ideal. The first group is comprised of synthesis, relation, continuity and discontinuity; the second group of identity, the relation of quality, negation and rationality; the third group of causality, totality and development; and the last one of values. The role of cognition is that of integrating experiences or phenomena, which Høffding called "items" and which are not produced by ourselves, with the forms of thought in order to bring about a maximum of continuity and rationality. But, as we will see later in detail, there are three domains where the completion of this task will forever remain unsuccessful. The final chapter concerns itself with ethics, religion and the understanding of life.

Høffding believed that human knowledge was not solely a result of the application of the forms of thought, and neither did he believe that it could be explained merely in terms of its utility with regard to practical matters. Høffding's theory of knowledge lies somewhere between Kant's notion of *a priori* categories and the theories of knowledge characteristic of pragmatism as they were developed by his contemporaries Charles S.Peirce and William James as well as Ernst Mach and James Clerk Maxwell.

His epistemological studies continued also after the publication of *Den menneskelige Tanke*, which was translated into both French and German, but unfortunately never into English, and his conclusions emerged in several of his minor works elaborating some of the categories. In 1914 he was elected as the first occupant of the Carlsberg honorary residence, which is a mansion donated by the founder of the Carlsberg Brewery for the use of the scientist most worthy of recognition. In the succeeding year he retired from his university chair. During the First World War Høffding worked for the Danish Red Cross and was its president from 1917 to 1921. In 1916 he published one of his most personal books, *Den store Humor* (Great humor), which expresses better than any other his attitude towards life.

To Høffding philosophy was related to what he felt strongly about, and the position he adopted in attempting to create unity and harmony between opposing views was very much a reflection of his own personality and character and of the inner conflicts of his youth. The prominent position Høffding occupied at this time in the world of Danish science and culture was not the result of calculated shrewdness, neither was it owing to elegancy of style revealing literary ability, nor owing to highly original ideas. Rather it was due to the far-reaching views and eclecticism of his philosophy, its integrity and level-headedness, and to his personal interest in all kinds of philosophical problems that he obtained recognition. As a result of his humanism and extensive knowledge, Høffding became and remained, throughout a long life, a person of high standing in the eyes of many. Through his published work and his teaching he contributed as none had done before or have done since to the consolidation of philosophy as a discipline among Danish scholars and scientists.

2. HØFFDING AND BOHR SENIOR

Niels Bohr's father, Christian Bohr (1855–1911), who was a physiologist, was one of Høffding's colleagues at the University of Copenhagen; both were members of the Royal Danish Academy of Sciences and Letters, and they were close friends, each admiring the other as a thinker and a creative scientist. In his *Memoirs* Høffding mentions that he and Bohr Senior, together with two other members of the Royal Academy, met after the sessions in each others' homes in turn. Here they discussed the topics which had been the subject of the evening's meeting. Høffding gives us the following report:

Now at some point in the period I have been talking about [around 1893] regular gatherings began to take place which I have taken and still take great pleasure in. It all started through my habit of joining the physiologist Christian Bohr after the meetings in the Royal Academy, often continuing my conversation with him in a cafe where we had supper. I had already been in touch with Bohr earlier, who at that time had learned that I was working on a treatise on psychology in which I intended to incorporate as many physiological views and results as possible. He offered his assistance by reading the relevant part. These evening visits which were thus initiated were followed by many more. In his and his wife's beautiful home in the stately residence of the professor of physiology I have spent many interesting and enjoyable evenings. Being a follower of the Leipzig scientist Ludwig, Bohr belonged as a physiologist to the movement which strongly insists on physical-chemical methods in physiology. ... Before long the physicist Christiansen was also one of the party at our cafe visits after the meetings in the Academy. ... The trio which was formed got tired of the cafe life, and so it was that we in turn went back to the home of one of us those Friday evenings when meetings in the Academy were held. Furthermore we gained a fourth member in person of the famous linguist Vilhelm Thomsen.[6]

The sessions in the Royal Academy took place every fortnight, 14 or 15 times during the winter, from the middle of October to the end of May. As a physiologist Bohr Senior was very interested in the methodology of biology and in the conflict between a mechanical and a teleological explanation of life, and most likely this problem was one the circle of friends discussed again and again.

The discussions between Christian Bohr, Høffding, Professor Christian Christiansen (1843–1917), later Niels' teacher in physics at the University, and Professor Vilhelm Thomsen (1842–1927) started when Bohr Junior was about 8 or 9 years old. When he was a little older he was probably given the opportunity to listen to the discussions every time they occurred in his home as a child. Niels Bohr has twice mentioned these meetings and what they meant to him. The first time was in his commemorative speech on Høffding made in the Royal Academy:

My first recollections of Høffding stem from some evening gatherings – described by himself in his Memoirs – when a small circle of scientists about a generation ago met regularly in each others' homes to discuss all kinds of questions which had caught their interest. The other members of this circle were close friends of Høffding's from their student years together, Christian Christiansen and Vilhelm Thomsen, plus my father who was a good deal younger, but whose friendship for Høffding over the years increased in intimacy. From the time that we were old enough to profit from listening to the conversations and until these gatherings in our home came to an end upon the early death of my father, we children were allowed to be present when the meetings were held at our house, and they left us with some of our earliest and deepest impressions. During the discussions, which were often very lively, Christiansen especially would tease Høffding in his typically good-natured way about philosophy's general aloofness from the world; but like everyone else present he was well aware of exactly to what extent Høffding's ability to understand and the impetus he had towards forming a general synthesis of differing points of view was, so to put it, the nourishing soil from which the ideas of the others sprouted, though marked by their diverse academic backgrounds and views of life.[7]

When his father died in 1911, Niels was 25 years old and, from what he tells us here, he apparently continued to take part in these meetings right up to that time. We also get the impression from this passage that Høffding was at the center of the circle.

A second allusion to these gatherings is to be found in his paper "Physical Science and the Problem of Light", written in 1957. In this paper he discusses the complementarity of mechanical and teleological considerations in biology. After quoting a lengthy passage from one of his father's papers in which he juxtaposes a teleological language with a mechanical one, Bohr Junior writes:

I have quoted these remarks which express the attitude in the circle in which I grew up and to whose discussions I listened in my youth, because they offer a suitable starting point for the investigation of the place of living organisms in the description of nature.[8]

When Christian Bohr died at the young age of 56 in 1911, Høffding gave a commemorative speech which was published in the journal *Tilskueren* (The Spectator):[9]

This man was carried off while at the height of his powers. Shattering and overwhelming was the news of his death. It was a happy death, indeed; he had just returned from work and went to the bed which was within a few hours to become his deathbed. To fall asleep in this way after a day's work is what many people could wish for. But in this case the words of the poet hold no value, "He fell asleep like the sun sets in fall". For it was not fall, not autumn, for him yet. He was not yet one of the veterans. He was as yet in the summer of life, still displaying in the searching ingenuity of his mind and the energy present in all his labor that he was first in rank among Danish scientists and with many years of fruitful work in prospect.

If it has fallen to my lot to speak here, it is not because I was a fellow-worker in his field. But I have followed the course of his work over a number of years with profit and admiration. He offered me a helping hand when I needed assistance for my own work from a specialist in his field, and it was this that first brought us closer together. Later I had the opportunity to follow his work over a number of years in the Royal Academy of Sciences and Letters. Whenever he began to speak, the evening became a festive one, all absorbed by the account of his new discoveries, the difficulties which he had overcome and the new problems which had emerged for him. For many years his research traced a single fixed line. In one of the most central areas of research, that of organic life, it was his task to investigate the boundary between life and the forces of inorganic nature – to discover whether the boundary was unmovable, and if so, where it was located. And for such an investigation the physiology of respiration was especially fitted. Bohr stands out as an independent figure in the era of the history of physiology which was founded by Johannes Müller and his disciples in Germany, and by Claude Bernard in France, and whose program was to trace the general effects of natural forces as far as possible in the processes of organic life. Bohr brought to light a provisional limitation of their action by showing that respiration was not entirely determined by the influence of external conditions. The enigma turned out to be located further back than the conflicting parties believed. There was a time when Bohr complained of his position being misunderstood, it being thought that he wanted to regress to so-called vitalism. But his opinion was merely that a new chain of investigation had to be established, especially on the influence of the nerve system on processes in the lung cells. His position here shows something of his character as a scientist, and in his last dissertation, published in the Festschrift of the University, he presents a clear exposition of his view. The premium he set on thought shows itself in his critical grasp of a range of problems, paying attention to the vast horizons which in the wake of every new chain of investigations are revealed to the true scientist – but at the same time with the firm conviction that only one road is given along which progress in the new areas can be made: unrelenting work, faithful to the spirit of stringent science. His vocation could have as a motto Lotze's dictum, "Only exhilarated by the great, but faithful to the small". He was for us all an instructive example, not only in virtue of the results his research brought forth, but more especially through his stature as a scientist.

He has done honor to his country. May our flag not be flown victoriously in the strife of the world powers, but may it be proudly flown above works of the mind, above achievements of thought, and may it be lowered at his bier.

I have been requested by the president of the Royal Academy and the vice-chancellor of the University to express our gratitude to our deceased colleague for everything he has meant to the scientific community at home and to the training of scientists. He was a colleague whose words had weight. His acute intelligence penetrated many obscure connections and his unfailing common sense was often salutary. The status of science at home was close to his heart. He had a sharp eye for the dangers which might menace within the limited horizons of a small country and for the rise of popularization which often lets undigested knowledge come to the fore as allegedly true knowledge. In his own untiring work he set a living example of combat against such dangers.

For those who were not colleagues but had the pleasure to be his intimates, his personality in private life displayed the same qualities as those which characterized him as a scientist. His free and vigorous mind often grappled with paradoxes, and he displayed his wit; facing his criticism, friends often could find themselves in a purgatory, but behind his criticism lay a respect for serious work and great fidelity in friendship. He was not of their number who allow themselves to be overwhelmed by feelings, or who offer or expect declarations of friendship, but such reticence gave force to the effect of a warm glance and his firm handshake. Neither was he of their number who make a habit of speaking about their philosophy of life and destiny. Deep within him lay a view which could be said to bear a resemblance to Goethe's lofty view of life, at once both naturalistic and idealistic, in which all that was petty and bitter faded into far horizons. There were few who knew their Goethe as he did. The lines of Mephestole were perhaps those he most often quoted; behind them lay an aspiration like that of Faust, only with the difference that he did not, unlike Faust, relinquish the work with flasks, levers and screws (Helbeln und Schrauben); the path to truth for him was not to be found outside the scientific laboratory.

Now we shall no longer see his bright eyes and fine, intellectual forehead which distinguished his sturdy figure; or listen to his instructive expositions or to his wit and sharp criticism.

His loss is a great one for science at home and abroad and for the circle of friends and colleagues. But we to whom he meant so much, however, turn now to the one and to those for whom he meant most. His personality pervaded his beautiful home through his life and his interests, and those to whom he opened it will understand how great the loss there must be felt. There he found love and understanding. In this as in other things he was a happy man. Not least in the pleasure he had of seeing young, capable persons develop in just those areas of scientific work in which he was especially interested. There had been every reason to believe that he was to have a happy working evening of life together with his ambitious sons and with the comfort of all the love which surrounded him at home. But it was not be so.

But from the thought of the suddenness of the loss and of all which the loss implies we return again to the memory of his personality and his stature as a scientist.

Honored be the memory of Christian Bohr.

The paper of Christian Bohr's which was published in the *Festschrift* of the University of Copenhagen, and to which Høffding refers, is the very same essay from which in 1957 Niels, approvingly, quoted a passage in the context of expressing the same view and which was also held by the small group to whom, as a young boy and as a student, he had had the opportunity to listen to. As Høffding emphasizes, it was Bohr Senior's aim, as a follower of Johannes Müller and Claude Bernard, to use mechanical ideas as far as possible in explaining the function of living organisms. He had no intention of conjuring up a non-physical entity, a soul, psyche, or *élan vitale*, as an explanation of the teleological features of the physiological processes. But instead of using only terms of external mechanical forces in explaining the function of the lung, the older Bohr claimed that a reference to the nerve processes was likewise necessary for an understanding of how the lung cells were working. However, Christian Bohr claimed furthermore, as did Høffding also (as I will substantiate in Chapter IV), that although many biological phenomena can be explained in terms of a mechanical description of the physiological processes, we also have a basic acquaintance with the nature of living organisms which cannot be described in a purely mechanical vocabulary.

It is difficult to say exactly when Bohr Senior became interested in the dispute between mechanism and vitalism in biology, for this was an issue which was in the forefront of much late-nineteenth-century scientific debate. Christian Bohr's conception of the heuristic role of a teleological description as necessary, along with the notion of a mechanical description in biology, may well have been formed before he met Høffding. David Favrholdt has pointed out to me that Bohr's father might have got this Kantian idea from Rasmus Nielsen when he attended Professor Nielsen's course in propaedeutic philosophy in 1872–1873. In the latter's book *Almindelig Videnskabslære i Grundtræk* (A general scientific methodology in outline), published in 1880, Rasmus Nielsen vehemently rejected the idea of a certain force of life or a specific matter of life on the grounds that it cannot be empirically confirmed. However, at the same time he also argues vigorously for the importance of the idea of teleology within biology because of the self-organizing character of any organic system. This creates a qualitative difference between the lifeless and the living.[10] Of course, it is not unlikely that there is a connection between these ideas and those of Bohr Senior, even though they were published several years after Bohr had passed his examination in 1873. But it appears from a letter from him to Høffding, dated 4th September 1900 and now in the Royal Library, that he had in general no high opinion of Rasmus Nielsen. Høffding had asked Christian Bohr to go through a paper on "Dansk Videnskab i det 19. Aarhundrede" (Danish Science in the 19th Century). Apparently, in this paper Høffding characterized Rasmus Nielsen, who also had been Høffding's teacher 10 years earlier, as "rousing". But when Bohr Senior made his comments on Høffding's paper, he had the following ironical remark: "... exclusively to unburden my heart, even though I would never try to let my opinion have any influence, I would like to say that if R. Nielsen, at the time I was listening to him, could be called "rousing", then he aroused only those students who could have continued sleeping without any loss, not the others". Certainly, Christian Bohr did not think that he had a tribute to pay, but he may still as a student have become acquainted with Nielsen's view of teleology. The same may be true for Høffding as well.

Both Bohr Senior and Høffding shared the view that inasmuch as purposefulness is experienced as a characteristic of physiological processes contributing to the maintenance of the organism, teleological descriptions cannot be eliminated from biology if all proper experience of the organism is to be accounted for. They did not mean by this that teleological descriptions set limits in principle to neuro-mechanical descriptions of biological phenomena. What they believed was that purposefulness is something which characterizes many physiological processes and that this holistic feature is essential to our experience of life. The description of the experience of an organism as a whole cannot be reduced to any mechanistic description, even if we are able to explain the working of biological processes by reference to a mechanical description. The purposive and mechanistic description, although in opposition to each other as regards their respective aims, do not really clash, since neither of them singly is sufficient and necessary for an exhaustive description of the organisms as perceived by us.

Looking back in 1957 on the work of his father and on the discussions, in the small phalanx headed by his father and Høffding, on the implementation of mechanical and teleological descriptions in biology, Bohr Junior undoubtedly saw this debate as a starting point for his reflection on complementarity. This suggestion appears very likely as soon as it is recalled that Høffding's position on these matters was based on a general position in epistemology, which he applied not only with respect to biology but also in psychology and other human sciences as well. It was by this broader outlook and firmer foundation that Bohr Junior was nourished when attending Høffding's lectures at the University.

Chapter II

1. THE EKLIPTIKA CIRCLE

Bohr was enrolled at the University in the summer of 1903 after passing the examination with which the program of "gymnasium" ("*studentereksamen*") culminates in the spring. As a freshman Bohr began studying physics under Professor Christian Christiansen, a study which continued until 1909 when he received his master of physics degree. Over the following two years he submitted and defended his doctoral dissertation at the University, and in 1911–12 he spent a year in England doing post-doctoral research in Cambridge and Manchester.

At the time at which Bohr was enrolled, all students were obliged to take a year-long course in philosophy at an introductory level, classes meeting six hours each week. This "propaedeutic philosophy" (or "*Filosofikum*" as it is called in Danish) was usually taken during the student's first year at the University. It has been generally assumed that the only training in philosophy Bohr received was this elementary course. However, I shall here try to establish the claim that Bohr also attended some of Høffding's more advanced lectures and seminars on various philosophical topics.

In the autumn of 1903 and the spring of 1904 Høffding taught the mandatory propaedeutic philosophy course which Bohr attended, and later on a group of students, who attended Høffding's public lectures and seminars, began to meet regularly after these classes in order to continue discussions of the various subjects. Among its members was Niels Bohr. In his book *Aarenes Høst. Erindringer fra mange Lande i urolige Tider* (Harvest of the years. Recollections from many countries in times of unrest), Peter Skov (1883–1967), former Danish Ambassador in Moscow, Ankara, Warsaw and Prague, writes about the group:[1]

After leaving school (1901) I began the study of law, partly in order to please my father, partly because I was attracted by so many other subjects, such as literature, art, philosophy, that I could not make up my mind which to choose. My studies allowed me plenty of leisure. For several years I attended Høffding's seminars. Here I met a group of friends who formed a small circle consisting of 12 members. Hence, the name "Ekliptika". Of their number were the Bohr brothers, the Nørlund brothers, Edgar Rubin, Brøndal and others who later became professors. *Quorum minima pars fui.* I have always associated with those from whom I could learn.

Another report on Ekliptika is one we owe to Vilhelm Slomann (1885–1967), an art historian, who conveys something of the atmosphere of the circle:

From 1905 onwards, Edgar Rubin, later a professor of psychology, organized several winters running, gatherings of fellow students of approximately the same age. At the beginning their number was 12. The meetings were of the type well-known in student circles: somebody read a paper, everybody has a cup of restaurant or boarding-house tea, and the amount of tobacco got through was prodigious. Natural science and social science, geography and the traditional humanistic disciplines – philosophy, literature, language, archaeology and history – were all represented, and the various concerns and approaches came to the fore or found expression in the exalted and sublime name: *Ekliptika*. Rubin continued to act as the vigilant and natural center of the circle, and most of the members became professors in due course, or had in some way or other their share in scientific work.[2]

Slomann also reports on the brothers Niels and Harald Bohr's role:

Niels and Harald were active participants at the meetings, and listened attentively to the various addresses and discussions. When the discussions were beginning to tail off, it often happened that one of them made a few general remarks about the lecture and continued in a low voice, at a great speed and with impassioned intensity, often only to be interrupted by his brother. Their mode of thought seemed to be co-ordinated; one improved on the other's or his own formulations, or defended in a heated yet at the same time good-humored way his choice of words. Ideas changed key and underwent refinement; there was no defense of preconceived opinions, but the whole of the discussion was spontaneous. This mode of thought *à deux* was so deeply ingrained in the brothers that nobody else could join in. The chairman used to put his pencil down quietly and let them get on with it; but when everybody moved in closer to them he might say ineffectually: "Louder please, Niels!".

Apart from the brothers Niels Bohr (1885–1962) and Harald Bohr (1887–1951), the brothers Niels Erik Nørlund (1885–1981) and Poul Nørlund (1888–1951), the former a mathematician and the latter a historian, and later Niels Bohr's brothers-in-law, Edgar Rubin (1886–1951), the philosopher and psychologist, Viggo Brøndal (1887–1942), the Romance-philologist, Peter Skov and Vilhelm Slomann were members of the circle. So were Lis Jacobsen (1882–1961), the etymologist, Astrid Lund (1881–1935), who in 1909 married Elias Lunding, a graduate from the Royal Veterinary and Agricultural College, Kaj Henriksen (1888–1940), the entomologist, and Einar Cohn (1885–1969), later Permanent Under-Secretary.[3] Gudmund Hatt (1884–1960), who was elected as professor of ethno-geography in 1929, seems also to have had contact with Ekliptika. There are several things that indicate such a connection. Some of the above-mentioned 12 members of Ekliptika cannot have joined the group until two years after its founding in 1905. Moreover, Slomann mentions that geography was represented in the circle. Hatt passed his examination for the "*studentereksamen*" in 1904, the same year as did several other members of Ekliptika, and the following year he went to America, where he remained until 1907. When back in Copenhagen he no doubt renewed his connection with the group. Furthermore, there is a sheet of paper filed in the Bohr Archive which has since been dated to 1915/16, on which somebody has drawn a hepta-pointed star, in each point of which a name and an address have been written. Five of the names belong to members of Ekliptika: Niels and Harald Bohr, Poul Nørlund, Viggo Brøndal and V. Slomann. The sixth is that of Elias Lunding and the seventh of Gudmund Hatt. An arrow around the star and a few words indicate the order in which the members were to arrange the meetings in turn; that is all. It no doubt represents what remained of the Ekliptika group who were still meeting regularly. The period during which these gatherings took place seems to have been around 1916, because Bohr was in Manchester from 1914 to 1916, while Slomann visited America in 1912–1914. The year 1916 would confirm the correctness of the addresses written on the star. And the date also fits in very well with the fact that N.E. Nørlund is not on it, since he was abroad in 1911–1912 and just after his return became a professor of mathematics at the University of Lund in Sweden. As for Hatt, his connection with the group has fallen into oblivion, due to the fact that he went to America just after the foundation of the Ekliptika, and perhaps because many years later he was accused by some of the members of being sympathetic towards the Germans during the Second World War. He was thereafter expelled from the scientific community.

That there was a connection between Hatt and the other members of Ekliptika, which I have already tried to substantiate in my paper "The Bohr-Høffding Relationship Reconsidered", is confirmed by letters from Rubin and Hatt to Høffding around that time. On 28th June 1906 Rubin sent him the following letter about Hatt:

Professor Høffding,

From a friend of mine, A.G. Hatt, I have received the enclosed letter which he asks me to pass on to you with a letter testifying to my acquaintance with him.

His father is a teacher in Holbæk and is, as far as I understand, a rather peculiar man who has brought Hatt up with an ardent enthusiasm for science. Both the peculiarity and the enthusiasm he has inherited; but his ultimate aim which is closest to his heart is to do something in art. He has had a very lonely childhood in an atmosphere marked by nervous tension.

I became acquainted with him here at the University, when we started as students here at the same time and met each other at the seminars on "Modern philosophers" given by the professor. ... He led a pitiable sort of life here in Copenhagen, and the disparity between what he had expected and what he in fact experienced, namely, an emotional upheaval and an unhappy love affair, was too much all at once for one who was never at ease anyway, and was the reason for his going to America about a year ago.

Here he has started on various things without any sense of fulfillment. (I assume that he has written about this himself). He now wants to try his luck at Harvard, and that, I believe, would be very good for him; "but it depends to a large extent on whether I can find a job which does not take up all my time". The people, whom he had spoken to, think, he writes, that an introduction to William James from yourself would prove useful to him.

Since I have faith in him and believe that he is made of the right stuff I hope that his request may be granted.

As far as I know, Professor Hans Olrik knows him and his family well, and takes an interest in him.

Yours sincerely,
Edgar Rubin

So it was Rubin who introduced Hatt to Høffding; partly because Hatt was a friend of Rubin, and partly because Rubin was the only one of the group for whom the study of philosophy was a full-time occupation, though Høffding himself was probably able to remember Hatt from his seminars, two hours per week, on modern philosophers at which only 19 students were present.[4] It was doubtless Rubin, as the person who started Ekliptika, who asked Hatt to join the meetings as Hatt was attending both Høffding's series of lectures (of which there were two) and his seminars in the spring of 1905, all on subjects which the other members of Ekliptika too must have felt attracted to. Hatt gives an account of his activities in his letter to Høffding.

Kansas Ind.Terr. 12th June, 1906

Dear Professor Høffding,

In the spring term of 1905 among your audience was a student of the name A.G. Hatt, who was also present at the seminars on modern philosophers. If, by chance, you, Professor, recall me that is splendid; otherwise my friend Rubin will certainly attest to the fact that I am speaking the truth. – In the month of August I went to America where I ultimately had the pleasure of making the acquaintance of a chemist, Dr. Sawyer, who took me on as his assistant, and I have used my spare time to learn.

At the very beginning of the letter Hatt makes it clear, by saying "among your audience was a student ... who was also present...", that in the spring of 1905 he must have attended both Høffding's seminars on modern philosophers and at least one of his public lectures on, as we shall see in a moment, free will or Kierkegaard. In what follows Hatt writes about his studies of the Indians in great detail, and relates that he is now visiting some of their territories. He then continues,

And finally, to what is on my mind; I have already taken up too much of your time.

It is my intention to go back to Boston with the aim of studying at Harvard. I have a great need to acquire thorough knowledge of something, since I know almost nothing, and in particular nothing in any comprehensive way. But for various reasons I have no wish to return to Copenhagen. However, there are some difficulties attached to my being admitted to Harvard. I have to pass an entrance examination which may well be demanding; but fortunately every individual case is judged on its own merits. The request I venture to make of you Professor, is that of some kind of letter of introduction to Professor James. Unfortunately you do not know me personally, but I have for a time been a student of yours. Dr. Sawyer thinks that such a letter would be very useful ....

Yours sincerely,
A.G. Hatt

Høffding met Hatt's request with a positive response, as we see from a long letter from Hatt, dated Dorchester Mass., 15th February 1907, in which he thanks him for a kind and encouraging letter. He makes an apology for not having written to Høffding earlier but he has often had Høffding in mind. He also relates that he has been admitted to Harvard, but that he has been able to take only one course and it was not one of Professor William James's. He complains of the expense of studying and of the fact that he has to work to earn a living. He then goes on to talk about his studies of the Indians, their poetry and mythology, and of the changes in their culture. In this connection Hatt adds, "Professor! What you write about the persistence of values through time is a lovely idea – but I don't believe it". However, Høffding was not a man to take offense at such a candid remark from a young student. Hatt then mentions the fact that he dislikes the Americans and refers to America as "the tomb of the nations". He is apparently very homesick. Finally, he returns to William James.

I liked professor James from the very first time I saw him. He was extremely kind. Unfortunately, it was not possible for me, as I have said, to study under him; but I have listened to some of his public lectures – it was especially interesting to listen to one on the pragmatic view of the concept of truth.

Hatt returned to Denmark during the summer of 1907, most probably because of the expense of studying at Harvard and because of his animosity towards the Americans. In a letter to Høffding, dated Holbæk, 26th November 1907, we learn from what Hatt writes, that he is once again attending some of Høffding's lectures, even though he had also started studying geography. This emerges from his apologies to Høffding, after having read a letter from him, for having mixed up the two days, Tuesday and Thursday, on which Høffding lectured, thereby mistaking the day on which Høffding was to hand back a manuscript of

Hatt's which he had gone through. He then thanks Høffding for the undeserved help he has received, adding that the sympathetic understanding he has met with him has been most helpful and profitable for an inexperienced young man. In another letter from Hatt around New Year, dated Holbæk 1st January 1908, he thanks Høffding for lending him two books which he now returns. He sends Høffding his best wishes:

> I wish you a happy New Year, dear Professor. Thank you for all the good I owe you. After I came home I have met with more than one pleasant surprise, and your friendship has been one of the most pleasant.

Such friendships connected Høffding, as we shall see, to several other members of Ekliptika, and in all cases the friendships continued until Høffding's death.

It was not only a common interest in philosophy which had brought various members of Ekliptika together. We have heard about the Bohr brothers and the Nørlund brothers, whose families were to be related through Niels Bohr's marriage to their sister, Margrethe Nørlund, later on. However, three of the other members were also related to each other. Lis Jacobsen, who was married in 1903, was a Rubin, and she was a cousin of Edgar Rubin. But so was Einar Cohn; his mother was their fathers' sister. Furthermore, N.E. Nørlund and Einar Cohn were class-mates at Sorø Academy, a very old and highly esteemed boarding school, and Edgar Rubin and Vilhelm Slomann took their "*studentereksamen*" at Slomann's school together in 1904. So a large number of the members of the circle were already friends or related before they started Høffding's seminars and set up their discussion groups.

It is not known exactly how many years the group continued to meet. Slomann tells us that "'Ekliptika' ceased to exist when its members were no longer students".[5] The meetings were started in order to discuss Høffding's seminars and lectures in philosophy, perhaps in the spring of 1905. Slomann has clearly stated that the gatherings of Ekliptika began in 1905. Rubin, who as we have heard was the prime mover of the circle, had, at any rate, by the summer of 1904 matriculated at the University, the same year in which Harald Bohr and Vilhelm Slomann did; but Høffding did not give a course in propaedeutic philosophy in the academic year 1904-05, as has been claimed.[6] He went on a trip to America for four months in the autumn of 1904.[7] So it is conceivable that Ekliptika was first established in the autumn of 1905. Brøndal, for instance, started his studies at the University in the summer of 1905. Moreover, both Kaj Henriksen and Poul Nørlund were not matriculated until 1906. Hence they cannot have been members of the group from the very beginning, since it is very unlikely that Ekliptika was founded as late as 1906 because Peter Skov (who took his "*studentereksamen*" in 1901) and Einar Cohn (who took it in 1903) both obtained their degree in 1907.

On the other hand, Ekliptika cannot have been started for the purpose of discussing the subjects of Høffding's propaedeutic lectures on philosophy, because Lis Jacobsen had already passed her examination in propaedeutic philosophy in the summer of 1901; Peter Skov in the summer of 1902; Astrid

Lund in the summer of 1903; Niels Bohr, Einar Cohn and N.E. Nørlund in the summer of 1904; and Harald Bohr, Edgar Rubin, Slomann and Hatt were all examined in the summer of 1905 by one of the other two professors of philosophy, Kristian Kroman, after having attended his lectures for a year. He had also examined both Einar Cohn and N.E. Nørlund the previous year as well as Astrid Lund a year prior to that. Viggo Brøndal passed his examination with Høffding as examiner in the summer of 1906; and Kaj Henriksen and Poul Nørlund passed their examination, also with Høffding as examiner, in the summer of 1907, as did Lis Jacobsen, Niels Bohr and Peter Skov before them.[8] In fact, we see that half of the members of Ekliptika were not examined by Høffding. From these considerations, I infer that members of Ekliptika apparently pursued philosophical studies after taking the examination in propaedeutic philosophy by attending Høffding's seminars and public lectures, i.e. *not* the introductory level "propaedeutic" survey course, on various philosophical topics in the following years, until most of them graduated between 1909 and 1912.

We know the topics of Høffding's seminars and public lectures held in these years from two sources:[9]

Autumn 1903:	1.	Propaedeutic philosophy.
	2.	The relation between religion and the state.
	3.	The problem of evaluation.
Spring 1904:	1.	Propaedeutic philosophy.
	2.	Kant's *Kritik der reinen Vernunft.*
Spring 1905:	1.	The psychology of free will.
	2.	Philosophical theories (drawn from the work of modern philosophers).
	3.	Lectures on Kierkegaard.
Autumn 1905:	1.	Propaedeutic philosophy.
	2.	Spinoza's *Ethica.*
Spring 1906:	1.	Propaedeutic philosophy.
	2.	Exposition of Henri Bergson's *Essai sur les données immédates de la conscience.*
	3.	Exposition of selected philosophical questions.
Autumn 1906:	1.	Propaedeutic philosophy.
	2.	Exposition of philosophical works.
Spring 1907:	1.	Propaedeutic philosophy.
	2.	Ethics.
	3.	Exposition of parts of the psychology and theory of knowledge.
Autumn 1907:	1.	The philosophy of religion.
	2.	The doctrine of the categories.
Spring 1908:	1.	Danish philosophers.
	2.	Kant's *Kritik der reinen Vernunft.*
Autumn 1908:	1.	Propaedeutic philosophy.
	2.	Introduction to Ethics.
	3.	Exposition of Høffding's own *Filosofiske Problemer* (Philosophical Problems).
Spring 1909:	1.	Propaedeutic philosophy.
	2.	Exposition of Leibniz's *Nouveaux Essais* and *Monadology.*

From this list we learn that the Ekliptika Circle was, as far as we can gather, established in the spring term of 1905 when Høffding was lecturing on Kierkegaard and the psychology of free will, one hour per week respectively, and giving the above-mentioned seminars on modern philosophers. This seems to be confirmed by the letters from Rubin and Hatt to Høffding. The lecture hall was always very crowded when Høffding entered. He was met by a hush of expectation from an audience of 300 people when he was going to talk about free will, of whom 100 were students, and of 140 people when the lecture was to be on Kierkegaard.

"The psychology of free will" was not only the title of a series of Høffding's public lectures, it was also the title of chapter seven of his *Psykologi i Omrids på Grundlag af Erfaringen* (An outline of Psychology on the Basis of Experience). There can be but very little doubt that Høffding referred to James in his lectures on free will and in his seminars on the theories of modern philosophers, having just returned from America where he had visited William James. In 1905 he also published a short paper "Begrebet Villie" (The concept of will), perhaps to serve as an introduction to his lectures, in which he mentions James by name attributing to him, among other philosophers, the view of the will as the fundamental faculty of the mind, the very same view Høffding had adopted himself in opposition to Humean tradition.[10] William James, in his *Principles of Psychology*, was the first philosopher to apply the word "complementarity" in the context of shifting attitudes in psychology, a phenomenon which much later became significant for Niels Bohr in his development of the concept of complementarity. There are also very good reasons for believing that Høffding's seminars on modern philosophers included an exposition of the philosophies of Charles Renouvier and Emile Boutreaux as well as of James Maxwell, Ernst Mach, Heinrich Hertz and Wilhelm Ostwald among others, as Høffding in 1904 had published a book on *Moderne Filosofer* (Modern philosophers) that contained a discussion of these and others philosophers and scientists. The others were Wilhelm Wundt, Roberto Ardigò, Francis Bradley, Alfred Fouillèe, Richard Avenarius, Jean Marie Guyau and Friedrich Nietzsche.

The date put forward above as being the likeliest candidate for the foundation of Ekliptika fits in very well with the fact that Hatt went to America in the second half of 1905. Both Peter Skov and Slomann state that they participated in the gatherings of the circle for several years; and Bohr must have done the same. Bohr himself has indirectly confirmed this in the last interview he gave: "In some way I took a great interest in philosophy in the years after my student examination. I came in close connection with Høffding".[11] Bohr seems, then, to have attended Høffding's lectures and seminars on Kierkegaard and on Spinoza's *Ethica* as well as his seminars on modern philosophers (which we know that both Hatt and Rubin attended); his lectures on his own philosophy of ethics, psychology and theory of knowledge; and maybe, too, his seminars on Kant's *Kritik der reinen Vernunft* in the spring of 1904, topics which have also influenced his later ideas on psycho-physical parallelism and the relation between the knowing subject and the object known.

If I am right in assuming that Ekliptika was established as early as in the spring of 1905, then at least three of those known to be members cannot have participated from the very beginning. As indicated by the name of the circle there were twelve of them. Gudmund Hatt, as we have seen, could have been one of the absent members. But today we do not have the least idea of who the others were. However, Viggo Brøndal may have joined the circle in the autumn of 1905, because he passed his "*studentereksamen*" in the summer of that year; and since Kaj Henriksen and Poul Nørlund only took theirs in the summer of 1906, these two students cannot have joined Ekliptika before the autumn of 1906. Unless, of course, they had started to attend meetings while still pupils at a "*gymnasium*", although this is very unlikely. At any rate they started at the University in the autumn of 1906 by attending Høffding's one-year course of propaedeutic philosophy. Here they must have met two brothers, Georg Cohn (1887–1956) and Naphtali Cohn (1888–1937), who both qualified as jurists in 1912. Although law students, they were also very interested in philosophy and attended Høffding's more advanced seminars as did the members of Ekliptika. Did they take part in the meetings of Ekliptika from that time on? It is difficult to tell because it seems that they were not related to Einar Cohn. However, the circle had most likely started to diminish in size at that time since both Peter Skov and Einar Cohn were preparing for their final examinations in 1907.

Although Georg and Naphtali Cohn were law students they each submitted a prize paper in philosophy, Georg on "Ethics and Sociology" and Naphtali on "The Relation between Locke's and Kant's Theory of Knowledge", and Høffding and the other members of the jury awarded both papers the gold medal. While still a student Georg Cohn also published a book on Plato's *Gorgias*. Shortly after Høffding died he wrote a long paper in the same journal in which Høffding had published his commemorative speech on the death of Bohr's father many years earlier. Here he portrayed Høffding and gave an account of Høffding's relations with the students during the period in which Ekliptika was flourishing:[12]

At the very beginning of their studies at the University of Copenhagen, all students have to take a one-year course in Philosophy. There are several professors offering courses and one is free to make one's own choice. The majority at that time chose Høffding. He was one of the great names in Danish scholarship, alongside the linguist Vilhelm Thomsen, perhaps the greatest. Being very philosophically inclined I not only followed these ordinary lectures but also his more advanced seminars with the students of philosophy. These seminars took place once or twice a week in the so-called Philosophical Laboratory or in Professor Høffding's home. He lived at that time in a beautiful old patrician building in [the street] "Strandgade" in [the part of Copenhagen called] "Christianshavn". Professor Høffding's flat was rather spacious but the ceiling was quite low. All the walls were lined with books, and the most recent philosophical literature was always to be found on his desk. He worked unrelentingly, and the open-mindedness and receptivity with which he always met any new and important ideas in the development of philosophy worldwide was characteristic of his personality and philosophy.

I regard the memory of these enjoyable and instructive gatherings in Professor Høffding's home as one of my best. Høffding's personality was extraordinarily charming. He had a funny face with a small snub-nose and two very lively eyes which with interest, expressing at the same time wonder and a certain dry humor, surveyed the wide world containing not only all the great philosophical puzzles, but the laughable aspects of things which prevented life from being too serious an affair. Høffding did not consider himself a handsome man, and in his *Memoirs* he describes very humorously how he while still a boy was brought to realize that he was not. For his father was a very handsome and stately man, and when Høffding on one occasion in a public baths told one of the other bathers who his father was, the person in question said: "Heavens, are you really a son of that handsome man?"

There was something Socratic about the whole of Høffding's appearance. With the greatest patience and with truly lively interest he listened to the questions of young students and to their early trials in coming to grips with philosophy. He let himself engage in discussions with them about their views and was able to awaken their philosophical interest and personal confidence. The students went to him as one they trusted when subject to doubts and scruples, for instance, with regard to religious questions, about which they may even have avoided talking to their close relatives. Professor Høffding has several times related such incidents and has also written about them in his *Memoirs*. It was at these more intimate gatherings that his talents as a teacher and one able to inspire others truly came to the fore. In his *Memoirs* he tells us that to him it was a source of great happiness to have been a teacher at the University. He appreciated the constant interaction between the study and the lecture hall. He further recalls the fact that where the students were concerned one generally found oneself touching on or talking about all that which constitutes one's main interest in life – truly the fact of having intercourse with young people at the age when their intellectual and spiritual foundations are being formed – all this makes life rich.

Bohr may certainly have experienced the atmosphere that Høffding created as described by Georg Cohn. Høffding was obviously very dedicated to his students. By listening to what students had to say, and by taking part in discussions with them, something quite exceptional at that time, Høffding created an interest in philosophy among all kinds of young students, an interest that stretched far beyond propaedeutic philosophy. Bohr was strongly drawn towards philosophy, and here Høffding played a pivotal role in directing and encouraging his interests in philosophy. Even before he enrolled at the University Bohr had met Høffding many times in the home of his childhood, where the discussion of philosophical questions fired his interest in the subject. When he grew older he no doubt got the opportunity to put his own questions to the participants at these gatherings. At the University, and perhaps already before that time, he read some of Høffding's many books and listened to his exposition of the great philosophers. Later on Høffding's influence was sustained through a close friendship which lasted for as long as Høffding lived.

But Bohr was not the only member of Ekliptika who was influenced by Høffding and remained in touch with him after their student years. Letters to Høffding, now kept in the Royal Library from several of Bohr's companions show that they were still meeting and corresponding with Høffding in the twenties. There are letters, besides those of the Bohr brothers, from Rubin, Hatt, Brøndal, Lis Jacobsen, Slomann and N.E. Nørlund. Some of these letters were sent to Høffding on the occasion of his seventieth and eightieth birthdays when it was impossible for the writer to congratulate Høffding in person.

One letter to Høffding on his seventieth birthday is from Brøndal, who graduated in 1912 and studied in France from 1912 to 1913.

Paris, rue Brisscade 6, 9th March 1913

Dear Professor Høffding,

The newspapers announce that you will be celebrating your birthday on Thursday and give an account of all that will be done to honor you and to pay tribute to you on that occasion.

Allow me too, though at present abroad and no longer a member of the group whose custom it was to meet to discuss philosophical questions with you, to send you my heartfelt congratulations.

What makes you the teacher to whom I shall continue to owe most – although the subject I chose was not yours – is the high standards of scholarship and the kind interest you have always shown young students.

Such standards have never repelled, nor was your interest ever narrow. One could confide in you about personal matters and projects which were remote from your own. This can be said of very few; there are only few to whom we, as young students, owe a debt of gratitude of such a degree. You have no equal as a teacher for the academic youth of Denmark.

I have no greater wish for you than that you for many years to come may continue to be vitalized by, and find pleasure in, intercourse with the coming generations of students that appear like blossom each year.

Yours sincerely,
Viggo Brøndal

Indeed, Brøndal's statement "at present abroad and no longer a member of the group whose custom it was to meet to discuss philosophical questions with you" may indicate that some of the members of Ekliptika were still gathering even though he himself was not a member any longer. But we have also seen that Brøndal's name figured on the star which, apparently, refers to a regular arrangement of meetings of several members of Ekliptika, probably around 1916. If this is correct, Brøndal must later on have been invited to join the remnant of the group once more. The sentence also seems to indicate that the Ekliptika group met with Høffding himself to discuss philosophy, a custom which Georg Cohn's recollections on Høffding seems to confirm. In 1928 Brøndal was appointed professor of Romance philology, and together with Louis Hjelmslev (1899–1965) he founded another internationally recognized "school of Copenhagen" in structural linguistics, as distinct from Niels Bohr's in physics.

The third well-known "school of Copenhagen", also having its source in Høffding's student circle, was established by Edgar Rubin in psychology. He graduated in philosophy in 1910. Between 1911 and 1914 he studied experimental psychology at Göttingen; thus it was not possible for him to congratulate Høffding personally on his seventieth birthday. Instead he sent him a letter expressing his admiration and sympathy.

Göttingen, March 1913

Dear Prof. Høffding,

I would like, with your permission, to be one of those who today congratulate you and express their thanks to you. You have been fired by a desire for earnestness and fidelity, by an impassioned impulse for your own person to come to grips with the truth about the existence of man, and through your extensive work in the service of this desire you have become a leader and mentor for people everywhere.

A typical Danish landscape induces no reason to believe in delightful things behind barren heights that narrow the wide view, rather there is a farm surrounded by field upon field, there is a village, and there is a stretch of meadow land, and there is a wood; there is something honest and free about it – it does not hide anything. So too is your thought, and so, indeed, is your very self. Only one thing have you concealed and, that is that you were born in 1843, and I hope for both your sake and ours that for many years this will be a fact of no account so far as your health is concerned.

I am glad to know that many people appreciate the value of a man who is plain and simple and who, since it is incompatible with his nature to know it himself, does not pride himself on his greatness; and I am glad to be one of many who today can send you their compliments in affection and admiration.

Yours
Edgar Rubin

Rubin became professor of experimental psychology in 1922, after having taught at the University as a reader in philosophy from 1918.

Rubin's experiments on the perception of figures and grounds and the conditions for their description attracted attention and won him recognition all over the world. For the purpose of the study of the relation between ground and figure in visual perception, he invented the famous Rubin vase in 1921. These studies resulted in a new understanding of the conditions under which the same stimuli brought about different states of awareness, the switch in perception from background to figure, and *vice versa*, and furthermore, under which conditions different stimuli afforded states of awareness of the same kind in form of color constancy, constancy of forms, constancy of size, etc. Thus perceptual phenomena gain, through his precise description, a fundamental psychological and epistemic interest, and his work on the psychology of perception, which was published for the first time in his doctoral dissertation in 1915, represents a continuation of Høffding's ideas of totality. The dissertation is regarded as the first major experimental work in psychology which clearly departs from "element psychology". Rubin's thought was that the mind forms a integrated totality of great complexity and that the various aspects of the mind must be regarded as wholes which cannot be described in isolation. This view of the mind stands in sharp contrast to the British tradition, being a clear extension of the Danish tradition as he had learned it through Høffding. Rubin himself seems to have reached many of his hypotheses through extensive epistemological studies based on Høffding's positivistic approach, apparently quite often independent of psychology. Not even gestalt psychology, a school with which he is often associated, was a serious starting point for his hypothesis. He placed himself outside any international school.

A final testimony of the admiration for Høffding from the Ekliptika circle, besides that of Bohr, to which we shall return, is one from the hand of Hatt. He graduated in 1912, and two years later he returned to America, spending a year studying at Columbia University. He made several expeditions to Lapland during the years after he had got his degree, and in the years 1922–24 he led an expedition to the Virgin Islands, which Denmark had sold to the United States shortly before. So he was not at home on Høffding's eightieth birthday. Being abroad he sent him a very long letter, of which only the beginning will be quoted here.

> St. Jan, 11th March 1923
>
> Dear Professor Høffding,
>
> My wife and I are sitting here beyond or at least on the outer fringes of the world having not been able to see the newspapers for a very long time. All the same it has occurred to us that surely it is today your birthday. Unfortunately, St. Jan has no telegraph and so we cannot send you our congratulations on the day. It would also be difficult to express them in telegraphic style.
>
> There are few people to whom I owe so much as to you. You have not only been my teacher – that you became long before I began studying at the University, for my father had many of your books – but it was particularly good for me to meet you personally, feeling that you wished me well and believed in me. This was stimulating. The splendid parting words you uttered on my setting out on my journey had a very stimulating effect and remain a source of encouragement for me. ...

Then follows a long description of Hatt's observations on the Virgin Islands.

Unfortunately for us, Niels Bohr seems to have been at home on all Høffding's red-letter days, so there exists no letter from his hand which could have informed us about his sentiments regarding Høffding. He undoubtedly was present in person to convey his good wishes on these festive occasions. When Høffding was to celebrate his eighty-fifth birthday in 1928, Niels Bohr, as we will see, made a public statement very similar to those of Brøndal, Rubin and Hatt. There is no question that most of the members of Ekliptika, if not all, were deeply influenced by Høffding's pattern of thinking during their student years, and that his ideas had a lasting impact on many of them. We are also able to ascertain from letters to Høffding that Niels Bohr, Slomann, Rubin, Lis Jacobsen, Hatt and Brøndal were all associated with Høffding later on.

2. BOHR AND WILLIAM JAMES

The fact which we have sought to establish – that Bohr not only attended Høffding's lectures on propaedeutic philosophy but also, later, his more advanced seminars and lectures on various philosophical topics – may help us settle the question as to when Bohr became acquainted with William James. As we saw in the Introduction, Bohr refers explicitly to William James in the interview given to Aage Petersen and Thomas S. Kuhn the day before he died, when he is asked by Kuhn whether he had read any of the classical philosophers. At the start he is uncertain and perhaps evasive until suddenly a distinct recollection of having read the chapter "The Stream of Thought" in *The Principle of Psychology* returns. To a further question from Kuhn, Bohr replies that he definitely read James before his time at Manchester in 1912, and probably around the time at which he was working on surface tension in 1905. It was during this year that the Ekliptika Circle was founded as a forum for discussions of Høffding's seminars and lectures, and it is highly imaginable that Bohr read parts of James's *Psychology* in connection with Høffding's public lectures on free will or the seminars on modern philosophers, both of which were given that spring after Høffding had just come back from a visit to James. James had clearly made a vivid impression on Høffding, and like him James was very involved with questions of free will and a basis for a science of psychology. What Bohr, in the interview, is excited about with respect to "The Stream of Thought" seems to be the claim James makes that any idea of a complex object is simple and psychologically indivisible; that is, although an object of thought may contain many elements, the thought itself does not consist of a manifold of ideas corresponding to these various elements but constitutes a simple unity.

In a previous chapter in his *Psychology*, however, James had proposed the use of the word "complementarity" for describing differences between a conscious and a subconscious self, with respect to the behavior of hysterical persons, for instance. James concludes,

> It must be admitted, therefore, that *in certain persons, at least, the total possible consciousness may be split into parts which coexist but mutually ignore each other*, and share the objects of knowledge between them. More remarkable still, they are *complementary*. Give an object to one of the consciousnesses, and by that fact you remove it from the other or others. Barring a certain common fund of information, like the command of language, etc., what the upper self knows the under self is ignorant of, and *vice versa*.[13]

These words may, of course, have fixed themselves in young Bohr's conscious or subconscious memory, but I am skeptical. I think this is a little coincidence that Jammer made too much of. "Complementarity" is a common term used in mathematics, for example.

However, Bohr's testimony of his acquaintance with James's writings has been dismissed as a lapse of memory. In a letter to Gerald Holton, Rosenfeld has "expressed his strong belief that the work of William James was not known to Niels Bohr until about 1932".[14] Rosenfeld recalls that in or about 1932 Bohr showed him a copy of James's *Principles of Psychology*, and he believes that Bohr had had a conversation with Rubin a few days before, which may be why Rubin sent the book to Bohr after their talk. Bohr showed excited interest in the book, and especially pointed out to Rosenfeld the passages on the "stream of consciousness". During the next few days Bohr shared his excitement with several visitors, and Rosenfeld retained the definite impression that this was Bohr's first acquaintance with William James's work.

Professor David Favrholdt has strongly defended what Rosenfeld is saying here, since he finds it very unlikely that Rosenfeld should be mistaken in these matters given his keen powers of psychological observation, his thorough grasp of the history of science, and his deep interest in Bohr's philosophy and its origins.[15] This may all be true. But it is also known that Rosenfeld was deeply attracted by Bohr's powerful personality, which has left a mark on so many who were in contact with him. So Rosenfeld may not be an altogether detached observer. Favrholdt, nevertheless, considers it to be evidence in favor of Rosenfeld's testimony that the lectures Bohr attended in philosophy were merely those forming the propaedeutic course, and that Høffding first visited James only after Bohr had passed his examination in propaedeutic philosophy, and thus had ceased to attend Høffding's lectures. Now, as I have shown above, we know that Bohr must have attended Høffding's lectures and seminars on more advanced philosophical topics for at least a year or two after Høffding had visited James as well as – at least – talking with Høffding when he met with the Ekliptika Circle after the seminars. Hence, Favrholdt has not produced circumstantial evidence that Rosenfeld's recollection is more reliable than Bohr's.

What we may do retrospectively, to be fair to Bohr's and Rosenfeld's seemingly contradictory statements, is to attempt to bring them into harmony. We know for sure that Høffding's friendship with James was known to Hatt, and that he only knew Høffding because he had attended Høffding's seminars on modern philosophers. Bohr, too, must have at least heard of James in Høffding's lectures on the psychology of free will and/or his seminars on modern philosophers in 1905, but he may not have read James's book until 1932. Since Høffding would have died the previous year, Bohr no longer had the opportunity to discuss psychological matters with him at a time when his interest in the application of complementarity outside the field of quantum mechanics was steadily growing. Quite naturally then, Bohr turned to his old friend Rubin about his interest in deepening his knowledge of psychology, and here Rubin probably recalled Bohr's attention to the work of James. Rubin must have been well acquainted with James's work already as a student of philosophy, and we can easily imagine James having being a subject of discussion in the Ekliptika Circle. Likewise, James's philosophy and psychology must also have been touched on many times at the Friday gatherings arranged by Bohr's father and Høffding, at which Niels was present, especially after Høffding's return from America. We also know that Høffding ran a seminar on James's *Principles of Psychology* in 1910. Of course, we have no reason to think that Bohr was present at that time, but Høffding must have referred to James very often in discussions with his three friends. So when Bohr was interviewed the day before he died, he may certainly have confused these two occasions, his becoming acquainted with James's psychology in his youth and his reading the passage about the stream of consciousness in 1932. I believe this to be a possible interpretation of Rosenfeld's testimonies in the light of Høffding's visit to James in the autumn of 1904 and his lectures and seminars in the spring of 1905. But since no additional evidence confirms Rosenfeld's testimony, I think it is most likely that Bohr had read at least some passages of James's *Psychology* at the time when he was a member of Ekliptika, as he has told us himself.

In that last interview Bohr also relates that already in his earliest years as a student he had a theory about free will and determinism as well as about the knowing subject. Thus, as we have seen, he believed there was an analogy between the solutions of multi-valued functions and those of the problems of free will and such matters.[16] As we shall see, it is reasonable to believe that his interest in these subjects sprang from Høffding's lectures on the psychology of free will, as well as from the conversations in the Bohr family home, where his father and Høffding very probably also discussed free will in the context of the debate concerning mechanism versus vitalism, and thus to believe that he must have been influenced by Høffding's thoughts on these matters and maybe by William James's as well. The problem in the case of James is merely that it is the very area of Høffding's views and interests which coincides with those of James that comes out in Bohr's comments, at least on psychology, free will, discontinuity and wholeness.

Here as elsewhere, however, we must bear in mind that although there are sound reasons for believing that Bohr was present at several of Høffding's seminars during his years as a student, he cannot have spent much additional time reading the philosophers themselves, simply because he was very preoccupied with physics – which after all was his subject. The lack of time to study any of the classical philosophers also explains why Bohr sometimes expressed himself philosophically in a misleading and inadequate way. Albeit very devoted to certain philosophical problems – which according to Høffding are the stuff of which philosophy is made – he had not enough spare time to acquire an exact philosophical vocabulary or delve deeply into the thought of any particular philosopher. Just before recalling once reading James, Bohr says, in answer to a direct question from Aage Petersen, that he knew Berkeley's ideas from Høffding's writings. Obviously, it is closer to the truth to say that Bohr became acquainted with the ideas of the philosophers through Høffding's, not so much because he had read their works himself as by the fact that their thoughts had first been ingested by his teacher. This claim is substantiated by Oscar Klein, one of Bohr's earliest assistants, who told Kuhn that Bohr "read, I think, some of Høffding's books, but I think he read very little more".[17] So what Bohr knew about philosophers and philosophical problems came to him through Høffding. By reading his books and attending his lectures he picked up a little bit here and a little bit there of other philosophers' ideas. But from Høffding's books and from the discussions with him Bohr became aware of what the important philosophical problems were and how Høffding thought they should be handled, as well as adopting Høffding's vocabulary for discussing and classifying these problems. A "moderately good" knowledge of other philosophers seems to be the right way to describe Bohr's knowledge of Kierkegaard as well.

3. BOHR AND KIERKEGAARD

We have it from a very reliable source that in the spring of 1905 Høffding gave a series of lectures on Søren Kierkegaard. If this is true, it is natural to suppose that Bohr attended them, an assumption I have argued for above. Central to Kierkegaard's philosophy is the notion of the incompatibility of thought and reality: we may create a system of ideas but not a system of reality because both the "existence", which is cognized, and the existing subject, which cognizes, are located in time, and only the singular or the individual exists in contrast to abstraction, which belongs to the realm of thought. So truth can, according to Kierkegaard, only be apprehended in a personal way: the truth is subjective just as subjectivity is truth. Kierkegaard saw his philosophy as opposed to the doctrines taught by the Romantic philosophers for whom all things could be subsumed under unity and continuity, and he called his thought "qualitative dialectic", thereby seeking to indicate the existence of sharply opposed qualitative distinctions.

We know for certain that Bohr once read Kierkegaard's *Stadier på livets vej* (Stages on Life's Way). In April 1909 Bohr stayed at Funen, one of the main islands of Denmark, working on the dissertation for his final examination. From here he sent his brother Harald a copy of Kierkegaard's book as a birthday gift with a letter saying:

> That is the only thing I have to send; nevertheless, I don't think that I could easily find anything better. In any case, I have enjoyed reading it very much, in fact, I think it is something of the finest I have ever read. Now I am looking forward to hearing your opinion of it.[18]

Some time later he wrote another letter expressing his less than full agreement with all of Kierkegaard's view:

> When you some day have read the "Stages", what you by no means must hurry with, you shall hear a little from me; for, I have written a few remarks about it (not in agreement with K.); but I do not intend to be so trite with my poor nonsense as to spoil the impression of so beautiful a book.[19]

We do not know which passages Bohr did not care for, neither do we know whether Bohr read any other book by Kierkegaard.

In "The Stages", which is hardly Kierkegaard's most philosophical book, Kierkegaard distinguishes between three life-stages or conceptions of life: the aesthetic, the ethical, and the religious stage. The aesthete seeks the pleasure which is the goal of all things. He has to keep himself suspended above the seriousness of life and not become too involved by withholding himself from, for instance, marriage and a permanent occupation. He lives solely for the moment as opposed to the ethicist, who exists continuously in time. He for his part regards responsibility and duty as essential elements of life. His relation to the opposite sex is realized in marriage with its obligations, and he finds his position in society. By contrast, the man of faith hopes for an eternal life which is realized in his relation to God. For him the merely humanistic standpoint of the ethicist is not enough. The experience of an intimate relationship with God is a condition of being truly alive. Kierkegaard lays great emphasis on the fact that these three stages are quite distinct and that one cannot pass from one to another without a "leap" in which one's attitude to life is radically changed. One must make a radical choice.

More than twenty years later, in 1933, Bohr had the opportunity to recall his reading of Kierkegaard during a visit to J. Rud Nielsen. Nielsen recounts in his paper "Memories of Niels Bohr", written in 1963, the incident as follows:

Knowing Bohr's interest in Kierkegaard, I mentioned to him the translation made by Professor Hollander of the University of Texas, and Bohr began to talk about Kierkegaard: "He made a powerful impression upon me when I wrote my dissertation in a parsonage in Funen, and I read his works night and day", he told me. "His honesty and willingness to think the problems through to their very limit is what is great. And his language is wonderful, often sublime. There is of course much in Kierkegaard that I cannot accept. I ascribe that to the time in which he lived. But I admire his intensity and perseverance, his analysis to the utmost limit, and the fact that through these qualities he turned misfortune and suffering into something good".[20]

Unfortunately, what Bohr here says is not very informative about the extent of his acquaintance with Kierkegaard's philosophy. But we may still be able to draw certain conclusions.

Bohr must have given an account of the impression that Kierkegaard had left on him earlier, either to Rud Nielsen himself or to others with whom Rud Nielsen had been in communication about it. However, Bohr's statement also seems to suggest that he had not read anything by Kierkegaard after 1909 and apparently not before. If Bohr had ever read more than one book by Kierkegaard why should he refer to only one very specific episode, the one we know of from other sources? Moreover, it is difficult to read Kierkegaard, it takes time to grasp his ideas, and one needs instruction to reap the full benefit of reading him. Consequently, if Bohr possessed any further knowledge of Kierkegaard than what he tells us of here, it would probably not have been gained from reading any other book by Kierkegaard but rather from having attended Høffding's lectures on Kierkegaard in the spring of 1905.

It has been suggested by Max Jammer, and Gerald Holton agrees and has pursued the idea, that Kierkegaard influenced Bohr in his work on the structure of the hydrogen atom in 1913, and later after 1926 in developing the idea of complementarity.[21] If this claim is justified, it would be more correct to say that Bohr has been influenced by Høffding's exposition and understanding of Kierkegaard's philosophy than by Kierkegaard himself. Apart from giving lectures in the spring of 1905 on Kierkegaard, Høffding had at that time published a monograph on Kierkegaard (1892), and in 1909 a book *Danske Filosofer* (Danish philosophers) in which he devoted a chapter to Kierkegaard, and in 1911 he wrote a paper on Kierkegaard and Nietzsche. As mentioned earlier, Høffding himself acknowledges his debt to Kierkegaard in his *Memoirs*, and in the field of ethics and psychology he saw himself partly as an exponent and a disciple of Kierkegaard. For instance, when Høffding writes, "As a psychologist Kierkegaard insists on continuity as far as he can; but as a moralist he postulates the leap, the jerk of the decision", he is expressing his own point of view just as much as Kierkegaard's.[22]

According to Høffding, Kierkegaard believed that the conditions to which the life of the spirit is subjected are quite different from those of natural life. Mental processes are characterized by discontinuity but natural processes by continuity, categories most essential to Høffding's own epistemology. Høffding has in his book *Den nyere Filosofis Historie* (A history of modern philosophy) given a very condensed account of his interpretation of these elements of Kierkegaard's thought. He characterizes an aspect of Kierkegaard's thought in the following way:

In Kierkegaard's ethics the qualitative dialectic appears partly in his conception of choice, of the decision of the will, partly in his doctrine of stages. He emphatically denies that there is any analogy between spiritual and organic development. No gradual development takes place within the spiritual sphere, such as might explain the transition from deliberation to decision, or from one conception of life (or "stadium") to another. Continuity would be broken in every such transition. As regards the choice, psychology is only able to point out possibilities and approximations, motives and preparations. The choice itself comes with a jerk, with a leap, in which something quite new (a new quality) is posited. Only in the world of possibilities is there continuity; in the world of reality decision always comes through a breach of continuity.

But, it might be asked, cannot this jerk or this leap itself be made an object of psychological observation? Kierkegaard's answer is not clear. He explains that the leap takes place between two moments, between two states, one of which is the last state in the world of possibilities, the other the first state in the world of reality. It would almost seem to follow from this that the leap itself cannot be observed. But then it would also follow that it takes place unconsciously – and the possibility of the unconscious continuity underlying the conscious antithesis is not excluded.[23]

This is a remarkable passage coming, as it does, from Høffding! For it certainly describes Kierkegaard "leap" in a way very similar to Bohr's description of the discontinuous change of atomic systems between stationary states as expressed in the quantum postulate.

Høffding discussed the same problem in some detail in his monograph on

Kierkegaard. Here he makes an incisive and critical comment on Kierkegaard's "leap" as it is displayed in *Begrebet Angst* (The concept of angst), since he thinks that Kierkegaard has not really understood the principle involved:

> It seems to be obvious that since the leap takes place between two states or between two moments no observation can be made of it. The description of it then becomes no description, since it can never be a phenomenon. What Kierkegaard aims to capture in his description slips through his fingers. He describes a succession of two moments or states (and what else can here be observed or described?), but what takes place between them he does not capture and it is in principle impossible for him to do so. And yet his assertions about it are as dogmatic as can be.[24]

Høffding adds that the dogmatic nature of Kierkegaard's preconditions, coupled with his use of images, blurs his perception of the core of the problem, viz. that the "leap" so described implies a will acting involuntarily.

Gerald Holton, too, has quoted the above passage taken from *A history of modern philosophy* in support of his claim concerning Kierkegaard's impact on Bohr. He rightly stresses that it would be absurd as well as unnecessary to try to demonstrate that Bohr had translated Kierkegaard's ideas directly and in detail into a physical context from their theological and philosophical context. However, he thinks that one should allow oneself the open-mindedness of reading Høffding and Kierkegaard

> through the eyes of a person who is primarily a physicist – struggling, as Bohr was, first with his 1912–1913 work on atomic models, and again in 1927, to "discover a certain coherence in the new ideas" while pondering the conflicting, paradoxical, unresolvable demands of classical physics and quantum physics which were the near-despair of most physicists of the time.[25]

I think Holton's proposal is well stated, and that there is, as we have seen, some evidence for his case. But I am not sure whether it is sufficient to warrant the conclusion.

In my Danish edition of Høffding's history of modern philosophy, which was a bequest from a late friend of mine, in his time a professor, there is the following marginal note in his hand against the above quotation: "Like 'quanta' in atomic physics". This comparison, which was made many years before anyone had even considered the existence of a connection between Kierkegaard and Bohr, is a piece of ingenuity. Nevertheless, pointing out the analogy when already acquainted with Bohr's theory of quantum jumps is one thing, but seeing an analogy between Kierkegaard's leap, such as Høffding had rendered it, and the movement of the electron in the hydrogen atom is quite another. In fact, what Bohr did, if Holton is right, was to draw an analogy where Kierkegaard definitively had denied the existence of any. For Kierkegaard only animate objects could undergo discontinuous changes of state.

However, contrary to Kierkegaard, Høffding did believe the "leap" to exist in inanimate nature. To support my last statement let me quote a passage from *Den menneskelige Tanke*, which was published in 1910 only a few years before Bohr launched his model of the atomic structure.

The ancient dictum that nature is in harmony with itself (*sibi consona* as Newton said) and makes no leaps often seems to be contradicted by observation. Indeed recent research yields counterexamples. In radium has been discovered matter whose emission of energy may seem to be at variance with the conservation of energy. In contrast to early Darwinism a spontaneous emergence of new types (by mutation) has been observed, generation which at present cannot be accounted for as being a result of successive processes.[26]

Høffding then adds that such discontinuities are also detectable in psychology and that William James in particular has pointed this out. On another page Høffding makes the following remarks:

Analogous problems may arise within a wide range of scientific domains. A problem similar to that with which radioactivity confronts physics is put to biology by mutations and to psychology by new mental formations and spontaneous changes in characters. And the attitude of our thinking with respect to such questions must be analogical too.[27]

According to Høffding, this implies that we have to treat all spontaneous processes alike, constantly investigating whether they really represent genuine discontinuities or not.

To conclude, I think that basically Bohr's knowledge of Kierkegaard was mostly second-hand, based on what he had read of him in Høffding's books and might have picked up in his lectures and seminars, but that this exposition might have stimulated further thinking in a young man interested in philosophical topics as much as problems in physics in the way Holton leads us to think was the case. Of course, whether or not Bohr, in forming his ideas of quantum jumps, found an analogy or a heuristic aid in Høffding's account of Kierkegaard's conception of a leap we cannot know for certain. I think, nonetheless, that Favrholdt's reasons for his dismissal of Kierkegaard's influence on Bohr are incorrect.[28] One of his arguments against such a thesis has it that Kierkegaard's philosophy is an extremely complex one, so it takes months to become acquainted with his universe and still more time to form a consistent interpretation of his ideas. But having such an interpretation would not be a necessary condition in the present case. One need not have, and most often has not, a comprehensive and critical grasp of a philosopher's work to be influenced by his ideas. One may have a rather superficial grasp of a man's philosophy and still find some leading idea or theme within it useful and inspiring. In fact, when it comes to "influences" one may certainly pick and choose different ideas from different sources, quite independently of whether one understands or agrees with any of these sources in their entirety, and regardless of whether they are consistent with each other. My contention is, however, that it seems plausible that Kierkegaard exerted an influence on Bohr only because Høffding on some issues, as some of those who have participated in the discussion have reported, shared the same ideas as Kierkegaard. The dichotomy between continuity and discontinuity, rationality and irrationality are features we also find as essential ingredients in Høffding's thought. It is by these ideas more than anything else that Bohr was inspired.

In the spring of 1905 Høffding lectured not only on the psychology of free will and on the philosophy of Søren Kierkegaard; that term he also gave a third series of lectures, seminars on modern philosophers, and we know for sure that at least two members of the Ekliptika Circle, Rubin and Hatt, were present. Since less than one year earlier Høffding had published a book with the same title as the above-mentioned lectures, it is reasonable to think that there was a close resemblance between them and his book in terms of content. If so, Høffding must have given an exposition of the philosophy of Charles Renouvier and Emile Boutreaux as representatives of the modern philosophy of discontinuity. Høffding must have felt a certain spiritual affinity with his two French contemporaries as he did to Kierkegaard, although he was only personally influenced by the latter. Professor Jammer has entertained the idea that the two French philosophers might have influenced Bohr. But Bohr never mentioned them, even though it is very likely that he had heard of them. However, their notion of discontinuities as real was quite familiar to Bohr through Høffding's own philosophy. Thus he might have noticed a certain similarity in the philosophy of all three, just as Renouvier's and Boutreaux's ideas might have been a subject for discussions in the Ekliptika Circle; but since his mentor held similar points of view and had a lasting intellectual connection with him, the influence of the ideas of the other two on Bohr must have been at best indirect, via Høffding; at the most Bohr might have regarded these ideas as "indicative of the winds of doctrine of the time", to borrow Henry Folse's expression, and hence they prepared his mind for what was to come.

4. THE STUDENT BOHR ON FORMAL LOGIC

Høffding used three of his own books as set books in his propaedeutic courses. There was a little textbook in the history of philosophy, *Kort Oversigt over den nyere Filosofis Historie* (Outline of the history of modern philosophy), published for the first time in 1898. Another was his widely known work on psychology, *Psykologi i Omrids paa Grundlag af Erfaring* (An outline of psychology on the basis of experience) from 1882, a book comprising more than four hundred pages. Both underwent several reprintings. The third book was a small treatise on logic, *Formel Logik* (Formal Logic). It was published in 1884, and reprinted in 1889, 1894, 1903, 1907 and 1913. When Høffding was Bohr's teacher in philosophy in the autumn of 1903 Bohr, as his student, had to read all three in order to be examined in philosophy the following year. But on becoming acquainted with the book on formal logic Bohr had some objections to it and apparently pointed them out to Høffding personally. This criticism must have made an impression on Høffding because when preparing the new edition of 1907, Høffding wrote to Bohr and asked for his comments. Formal logic had never been Høffding's strongest point but it tells us something about Høffding's open-mindedness that he approached a mere student for criticism. But it also tells us that, even at this early stage, he must have had high regard for the young Bohr's intelligence.

At the beginning of the last interview he gave, Bohr mentioned this episode;

and earlier both Høffding and Bohr made public statements about the incident. In the preface of the new edition Høffding thanks "one who was once a member of my audience" for critical remarks he had benefitted from. This expression seems to indicate that Bohr was no longer attending any of Høffding's seminars at the time the preface was written, around New Year 1906–1907. But Høffding may also be referring to the fact that, after Bohr had passed his "*Filosofikum*", he would no longer be in the audience of the course in propaedeutic philosophy where this was the set book, even though from time to time Bohr might have attended Høffding's seminars.

Bohr had, at a much earlier time than of the last interview, mentioned the episode in his memorial speech on Høffding in the Royal Academy:

> As for myself, I first came directly into contact with him in a scholarly way through some discussions on a philosophical topic which, given Høffding's philosophical approach and inclinations, might be said to be confined to the periphery of his interests, namely formal logic. Although Høffding did not himself attach any great importance to it as one of his works, his brief exposition of logic which he used in his lectures is an instructive book in which, as was typical for his way of thinking, general psychological experience form the vivid background, even for discussions of the classification of logical propositions. For a young man whose main interest was the mathematical treatment of the problems of natural science, questions of stringency of systematization were very much to the fore. I will never forget some evenings more than 25 years ago when I was allowed to visit Høffding in his home, at that time in Strandgade, and discuss these questions, and that he, a scholar so familiar with the history of thought and its infinitely varied aspects, with indescribable kindness and patience listened in order to discover whether in the young student's observations, colored more by enthusiasm than by clarity, there might be found the least scrap of anything new for him to learn in scientific and pedagogic matters. Høffding's unique impartiality on such occasions was to a large extent the reason for the influence he exercised and the encouragement in independent reflection he could give, and one would be the recipient of all this, and be almost unaware of it.

From what Bohr here tells us one might draw the conclusion that he had never taken part in a discussion with Høffding in his parent's home before beginning his studies at the University. This may be true. When Høffding, Christiansen and Thomsen visited his father he must have spent most of the time merely listening to their discussions, being only occasionally allowed to put a question or to make a brief comment. For at that time it was very unusual for children to be allowed to participate in adult discussion. Bohr also mentions that it was over 25 years earlier that he visited Høffding several times in Strandgade to discuss his objections to Høffding's exposition of formal logic. If what Bohr remembers is true, these visits must have occurred earlier than 1906. Most likely they took place as early as in 1903–1904 when Bohr was attending Høffding's course in propaedeutic philosophy and had to read his *Formel Logik* in preparation for the examination. This harmonizes with the fact that two years later in 1906 Høffding asked Bohr, in correspondence, to go through the preprints of the new edition.

Five letters from Høffding to Bohr at the turn of the year 1906–1907, kept in the Bohr Archive, give us more information concerning the substance of Bohr's objection. In the first letter Høffding asks for Bohr's comments on his book on logic as a revised edition was in preparation. Since Høffding goes straight to the point in the letter he seems already to have discussed this matter with Bohr.[29]

26 C. Strandgade, 22/11/06

Dear student Bohr,

I hereby send you the first proof sheet of the new edition of "Formel Logik" and ask you to peruse it with your usual critical eye. Your comments, if any, may be written on a piece of paper and enclosed.

Yours sincerely,
Harald Høffding

Bohr must have sent Høffding his criticism promptly, because nine days later he received another proof sheet.[30]

1/12/06

Dear student Bohr,

Hereby the second proof sheet. Will you send it to me in an envelop when you have read it and give me your comments on an enclosed sheet of paper. Will you take special note of the principle of duality (p.27).

Yours sincerely,
Harald Høffding

Here we learn that Bohr's main criticism has something to do with the principle of duality, which is what Høffding, following Jevons, calls the principle of excluded middle.

In the 2nd edition of *Formel Logik* Høffding informs the reader that he accepts "the logic of identity" or "intentional logic" developed by Leibniz, Boole and Jevons as the foundation of his outline of logic. A more detailed presentation of this view is given in the treatise, "Det psykologiske Grundlag for logiske Domme" (The psychological foundations of logical propositions), written in 1899.[31] In the 4th edition of *Formel Logik* from 1903, the one Bohr read for the "*Filosofikum*", Høffding formulates the three logical principles as propositions concerning "concepts" or classes: 1. The principle of identity, $A = A$, means that a well-defined "concept" or a well-circumscribed class is identical with itself; 2. The principle of contradiction, $Aa = 0$, means that the logical product of A and its negation a is a "concept" void of any meaning or an empty class; 3. The principle of duality or the principle of excluded middle is defined as follows: "A concept (B) must either contain another concept (A) or its negation (a). It can be expressed such: $B = BA/Ba$, as the sign / expresses a mutual relation of exclusion".[32] Høffding rejects a formulation of the principle of duality as $B = A/a$ owing to the alleged fact that if $B = A$ does not hold, then $B = a$ does not hold automatically, since B may only be partially identical with a. He writes further, "According to the principle of duality we have to choose between two possibilities: either to relate or not to relate a concept to a different concept as a stricter determination of the latter". And finally he adds that "the principle of duality includes the principle of contradiction, because if B at the same time could be combined with both A and a, then Aa would be a valid concept". Evidently, Høffding has neither understood the principle of duality, nor has he given it a consistent formulation. One might have expected Høffding to have defined the principle in accordance with a modified Leibniz-Boole-Jevons notation as a logical sum or disjunction of A and a equalled to the logical universe: "$A + a = 1$". If he had done that he would not have gone astray.

The assimilation of the principle of contradiction to the principle of duality seems to have aroused the suspicion of the young student of physics. Judging from Høffding's subsequent letters, we can infer that Bohr apparently argued that it is quite possible for a concept to contain both another concept and its negation. But Bohr's own suggestion of a formulation of the principle of duality, as revealed by Høffding, is not an improvement on Høffding's.[33]

4th Dec. 1906

Dear student Bohr,

I still have some scruples concerning your suggestion on the formulation of the principle of duality.

According to your suggestion it should be formulated as such

$$CA = CAB$$

$$cA = cAb$$

But why only take the two cases CA and cA? Is it not precisely the old formulation ($A = AC/Ac$) that is being presupposed here? Will not every explanation of why we are using this formulation provoke the question why we only speak of CA and cA? This appears very clearly as we take your formulation [illegible] for contraposition:

Having $A = AN$

we may put $Cb = CbA$, $cb = cba$, and since $Cb = CbAN = 0$, then $cb = cba$ may hold. We assume here without further proof that cb holds whenever Cb is invalid. Isn't this the old difficulty once more? To my mind this can only be a move into a *regressus in infinitum*.

Apart from this I believe that generally it will be simplest to keep to the old formulation – with the reservations you have shown to be necessary and which we must of course keep in mind. I am very grateful to you for your valuable help, it is of great interest for me to discuss these things with you. I do hope that I don't take too much of your time by asking you to express your opinion about the doubts I have mentioned.

Yours sincerely,
Harald Høffding

As far as it is possible to understand Høffding's presentation of and objections to Bohr's formulation, he seems to be right in his objections. Bohr's definition of the principle of duality certainly runs into an infinite regress, and this is possibly the reason why he is unable to get Høffding to join him in agreeing that the two principles in question are logically distinct. The last interpretation finds support in the following letter.[34]

6th Dec. 1906

Dear student Bohr,

I would appreciate it if you would run through the enclosed proof sheet which is the last. Pay special note to p.37. You will see from my corrections on which formulations of the principle of duality I have settled on. Since emphasis must be laid on the fact that the letters denote *intention*, and that the sense of the principle of duality is such that there is a limited choice between the two *combinations AB* and *Ab*, I have abandoned the idea of formulating it as an equation and merely write that it may be put in this way: *AB*/*Ab* (where / stands for a mutual relation of exclusion). In what follows I state that the principle of duality expresses a relation between two combinations (not, as written earlier, between two propositions). No reservation is needed now. Because the distinctive features, which hold for the type of vertebrates, either have to occur together with the distinctive features which hold for the class of mammals, or together with such which do not hold for this class. This formulation now agrees with both higher and lower concepts.

With respect to the use of the principle of duality, the new formulation fits very well with the making of the combination in connection with indirect inference. And [illegible] for immediate inference (p. 37) may also be represented clearly, since here it merely concerns getting rid of one of two possible combinations.

Your letter did not remove my doubts. If one makes a clear-cut distinction between the principle of contradiction and of duality, the latter is certainly *presupposed* whenever one assumes that it is not necessary to consider other combinations than *AC* and *Ac*. Now, I would like to hear, when the enclosed sheet is returned, whether you have objections to the new formulation.

Yours sincerely,
Harald Høffding

This letter from Høffding indicates that Bohr has rightly argued for a distinction between the principle of contradiction and the principle of duality but that he still has not given the correct formulation of the latter principle.

Let us now turn to what Høffding actually writes in the revised 5th edition about the principle of duality. He begins paragraph 27 on page 27 (in the letter he wrongly writes 37) by saying:

Whereas I cannot combine *A* with *a* I can combine it with a different concept. Concerning any such concept I have consequently the choice between two possibilities: to combine *A* with this concept (for instance *B*) or not to combine *A* with it. There are only these two possibilities; we are consequently faced with a contradictory relation. This is expressed in the principle of duality: The concept *A* must either be combined with *B* or not be combined with *B*.[35]

However, this is very similar to what Høffding wrote in the 4th edition. It is also clear that Høffding has gained nothing by talking about combinations instead of propositions, and that he only makes the things worse by his attempts to incorporate several concepts at once into the definition of the principle.

Høffding then comes forward with the example of the vertebrates: they have either to be mammals or not to be mammals, something which he had not mentioned in the earlier versions but which may be an attempt to meet Bohr's criticism. The example seems to guide Høffding to the apparently right formulation when, a little later, he writes, "Every concept and its negation divide the world into two parts (*B* and *b*)". This is as close to a consistent definition as he could wish for. But obviously he is not able to grasp the scope of what he says here. Neither is Bohr, so it seems.

Anyway, Høffding appreciated Bohr's criticism, and expresses this in a letter after hearing that he is to be awarded the gold medal for his prize paper to the Royal Academy.[36]

Friday night, 25th January 1907

Dear Student Niels Bohr,

It was a great pleasure for me this evening in the Royal Academy to hear that your dissertation has been awarded a prize. I congratulate you on the splendid result you have achieved at your young age and take the opportunity to thank you for your valuable cooperation.

Yours sincerely,
Harald Høffding

After the discussion of formal logic Bohr continued to see Høffding at least three or four times every winter when Høffding visited the home Bohr still shared with his parents when the Friday meetings were held here. When writing in his *Memoirs* about his move around 1907 from Strandgade to his new home at Carl Berhards Vej Høffding comments, "I had sufficient room for the gatherings which I held now and then at home with the participants of my laboratory colloquia. Also the pleasant Friday meetings with Vilhelm Thomsen, Christiansen and Bohr went on". So even if Bohr no longer attended Høffding's seminars and colloquia, there was still an opportunity for both meetings and intellectual contact between Bohr and Høffding until 1911, when Christian Bohr died, and the connection eventually developed into a friendship that lasted until Høffding's death.

Chapter III

1. THE FRIENDSHIP

After Bohr had stopped attending Høffding's seminars and his participation in the meetings in the home of his childhood had ceased, he and Høffding were still in touch. The Bohr Archive has in its possession a letter from Høffding, dated 14.05.1911, in which he sends his congratulations on the occasion of Bohr's defending his doctoral dissertation.[1] Later, in the same year, which was marked by his father's death, Niels Bohr went to England. He arrived at Cambridge at the end of September 1911, where he stayed until the beginning of April 1912. He brought along letters of introduction from Høffding to various people at Cambridge. On a postcard, dated 16.10.11, now kept in the Royal Library of Copenhagen, he thanks Høffding for these letters:

16–10–11 Eltisley Avenue 10,

Dear Prof. Høffding,

Many thanks for your letters of introduction, which I have much appreciated. I have visited Prof. Sorley and Sir. G. Darwin and also Miss Jones in Girton College (to whom I have remembered to convey your thanks for her book), and they all have been so extraordinarily friendly towards me and have asked me to send you their kindest regards. Being here agrees with me and I have met many people who have been stimulating to talk to. I've had to set myself up as a student and must, for instance, as member of Trinity College wear a long black gown and flat black cap, whenever I put in an appearance at the University, and also when I walk in the street after dark. The whole setting is indeed new and very strange, but I think that there is an atmosphere about it all which is so appealing, something which I already understood from you that you experienced when you were here.

With kindest regards,
Yours sincerely Niels Bohr

From Cambridge Bohr moved on to Manchester to work in Rutherford's laboratory, returning to Copenhagen at the end of July 1912 to be married. Back in Copenhagen he worked on the structure of atoms and their spectra, publishing his famous results during the year 1913. In 1914 he travelled to Manchester once more, staying there for the next two years. But in 1916 he went back home again when he was nominated professor of theoretical physics at his former university.

There is no tangible evidence today of the connection between Høffding and Bohr during the years 1911–1922. However, one letter dated 1920 from Niels's brother Harald Bohr to Høffding has survived and is kept in the Royal Library. In 1917 Bohr became a member of the Royal Academy, where of course Høffding and Bohr would meet. Still, it seems reasonable to assume that they were also in close contact with each other during these years. An undated letter to Høffding from Bohr, now in the Royal Library, in which Bohr apologizes for all the trouble a telephone call of his had caused – a call in the course of which he had asked for Høffding's help to arrange a meeting – must have been written between 1920 and 1925.[2] Three letters from 1922, two from Høffding to Bohr and one from Bohr to Høffding, have survived and are in the Bohr Archive. Here too are seven letters from Høffding to Bohr written between 1928 and 1931 and two letters from Bohr to Høffding. In Bohr's private correspondence, as distinct from his scientific correspondence, there are four more letters or lettercards written by Høffding to Bohr between 1922 and 1927, one dated 10th November 1922 and in which Høffding congratulates Bohr on being awarded the Nobel prize. In another of these, dated 3rd February 1925, Høffding closes by thanking Bohr for his visit the previous evening which had given great pleasure to both Høffding and his wife. In addition there exist two further letters from Bohr to Høffding, now in the Royal Library, dating from this period. In one of these, dated 30.12.1928, Bohr says something which also indicates that they had been in touch many times during that period:

Hornbæk, 30.12.1928

Dear Professor Høffding,

As the year draws to its close I feel an urge to give expression to all the joy and encouragement I have received on my brief visits to your home at Carlsberg. Apart from what I always learn from our conversations, I feel, as I have no doubt often said before, that it is like entering a world elevated above that of everyday life and which is closely connected to the memory of my father and Professor Christiansen. Not least in that context does my mind also turn to the great experience the appearance of your "Memoirs" in the year that's past has been for our entire circle. ...

Yours sincerely,
Niels Bohr

These visits are also mentioned by Høffding in his correspondence with Emile Meyerson. This is written in French and was published some years after Høffding's death.[3] The correspondence began with a letter of 25th February 1918 from Meyerson and continued over the next thirteen years until Høffding died. Bohr is mentioned nine times in Høffding's letters, first in a letter dated 20th May 1923. The next time is in a letter of 12th February 1924 where he refers to Bohr as "mon ami Niels Bohr". A letter of 13th August 1928 tells more about their friendship and discussions:

I have had most interesting talks with Mr. Niels Bohr especially on the irrationality brought into physics by the theory of quantum mechanics. Mr. Niels Bohr is not only a great physicist, he is also interested in philosophy and literature, and he surrounds his friends with warm sympathy. Some months ago I was ill and in bed for several days, and I think that my wife had spoken about my illness as being a most serious affliction; Mr. Bohr came over on several evenings spending hours sitting at my bedside. He talked to me about his works and about other interesting subjects, and he read to me some works by one of his favorite poets, while Mrs. Bohr was entertained by my wife. It was a great comfort to us in our uniform and isolated lives.

In a letter of 3rd October 1929 Høffding mentions that his wife is in hospital and that he is depressed on account of this. In this connection he again refers to Bohr and says:

However, what always gives me great pleasure are the visits of Mr. Niels Bohr. He informs me about his works *and we discuss their philosophical consequences.* (My emphasis)

And in the following letter to Høffding, dated 11th October 1929, Meyerson replies:

What you tell me about the visits of Mr. Bohr has pleased me greatly but not in the least surprised me; really superior men like him generally have a good character, and it is understandable that, his scientific work having led him to reflect on philosophy, he gets immense pleasure from, and finds great profit in, talking with you.

That Høffding and Bohr had many discussions about science and philosophy around that time is also testified to by a letter to Magrethe Bohr from Greta Høffding, who wrote from the hospital on 6th March 1930.[4]

I remember with pleasure (now also with sadness) the many pleasant evenings spent at your home while your husband and brother-in-law and my husband discussed the great questions of science at Dantes' Square [The Royal Academy of Sciences and Letters] – only to return afterwards in high spirits, everyone talking all at once – though not echoing each other.

When Høffding had published his *Memoirs,* in which he portrays Bohr Senior, he not only received a letter from Niels expressing his gratitude but also from Ellen Bohr, Niels's mother, and his brother Harald. From Ellen Bohr's letter, now at the Royal Library, we learn of Høffding's intellectual influence on Niels and their friendship:

Ahlefeltsgade 18 st, 29/4

Dear Professor Høffding,

Your exquisite and appreciative portrait of my husband has touched me deeply, prompting heartfelt gratitude towards you who have erected him such a beautiful memorial. I would also like to tell you how happy I am for what you are to Niels, and I also give you my heartfelt thanks for it. Harald and I have often talked about how happy and grateful we were because of your having transferred the friendship you had for my husband to him.

Sending kind regards to your wife, I remain, yours faithfully, Ellen Bohr

Ellen Bohr's words to Høffding are of much interest. I do not believe that the phrases can be dismissed as mere kindness and courtesy. To a native speaker of Danish such as myself, her words have an intensity that goes beyond that. And if she did not feel gratitude towards Høffding for what he did for her son, why should she mention that she *and* Harald had talked about Niels's relationship to Høffding several times?

The day after Høffding had received the letter from Ellen Bohr, he got another from her other son, Harald, who also expressed his gratitude for the things Høffding had said in the *Memoirs* about his father. After quoting a line of a Danish poet, a line which was once included in a letter to Høffding and which Høffding presented in his *Memoirs*, Harald closes his letter with the following tribute:

I have felt so strongly that this togetherness between people like Henriques, Hjelmslev, which Niels and I depend upon, cannot be more aptly expressed; and though we never get to doing so, I beg permission to say that I hope you have been aware of the esteem in which you are held in our circle – and there is no knowing how great the circles are where the same applies. I hope that you will not be offended or find it forward of me that I have allowed myself to write these lines to you as the expression of a deeply felt thanks from one of the many, many who – even though in my own case it has been for only a brief period – have had the happiness of making your acquaintance and for whom it has had a significance, we shall never forget.

Although these words may seem a trifle sentimental they are sincerely meant; and from what we have seen here, there can be little doubt that both Harald and, especially, Niels Bohr were very fond of Høffding.

As was mentioned earlier, Høffding got married for the second time in 1924 to Greta Ellstam, a woman much younger than himself. Unfortunately, her neurosis developed into a mental illness, which circumstance took its toll on Høffding during his last years. And he suffered a great blow when in 1930 he bore her to her grave. When it occurred, Niels Bohr expressed his condolences to Høffding in the following letter, now in the Royal Library.

Tisvilde, 19–8–1930

Dear Professor Høffding,

My wife and I were greatly saddened by the news of your wife's death, and we should both of us like to express our heartfelt sympathy with you in your great grief. We will always treasure the memory of your wife's warm, enthusiastic and richly gifted personality which you in your Memoirs have paid tribute to so beautifully. From the days that were difficult too, when illness taxed her constitution and clouded her mind, the memory of her struggle and efforts abounds with valuable instruction for Margrethe and I. The memory of those years, sad ones for you, are for both of us, however, colored by our admiration of the ideal you have represented and still do, grateful as we are for the example of human worth you have given us. Though I never will be able to find words to express my feelings, I believe that you have some impression of how often I have been moved as you told me about your life, which has been so rich in both external and internal events. I have never felt how the condition of human life with its equal amount of form and content determine harmony, stronger than I do now. I know how poor words are, but I would very much like to express our sympathy and my gratitude for all that you have meant to me, you who were to strengthen the bond between past and future in a very special way for me.

Yours faithfully,
Niels Bohr

A year after the death of his wife, Høffding himself died: a year which for the first time in his adult life was quite unproductive, but he was still in touch with members of *Ekliptika*. In 1931 as Rubin was planning the International Congress of Psychology in Copenhagen he asked Høffding, on the behalf of the executive board, to be its president, but Høffding refused. He had by then become too old and too tired.

So far as I know, the various statements and letters which have been examined here are all the evidence we have for the claims that Niels Bohr not only received his general philosophical training from Harald Høffding, but also that Høffding exercised a very strong personal, as well as intellectual, influence on Bohr, and that they became friends. I find the evidence sufficient to conclude that Bohr and Høffding were intimates and that Bohr owed Høffding much. Our next problem is, indeed, that of determining the form and the pervasiveness of the influence.

2. THE YEARS AROUND 1927

The meetings between Bohr and Høffding seem to have become more frequent around 1927, the year in which Bohr first presented the idea of complementarity. In his New Year letter of 1928 Bohr expresses his appreciation of the brief visits he had paid Høffding and, as we shall see, he also refers to these visits in a letter to Dr. Sorainen in 1946, pointing out that "as it can been seen from Høffding's latest work I had, in the years before his death, many deep and searching discussions with Høffding on recent developments in physics". As Bohr mentions here, Høffding had referred to these discussions in his paper "Psychology and Autobiography".

While I am claiming that Høffding's philosophy had a great impact on Bohr, I do not intend to suggest that this influence was merely a result of these discussions. Already from his youth, as Høffding's student at the University and from their conversations later on, Bohr was acquainted with and had absorbed Høffding's philosophy, so that it had, by that time, become a part of his own intellectual baggage and philosophical outlook. I think this is evident from the public testimony which Bohr gave on three occasions as well as from all the points I have made in the previous sections. But what seems to have happened in 1927 and 1928 was that Høffding directly drew Bohr's attention to the existence of the observational standpoint in psychology, similar to that in quantum mechanics, both relevant to the general problem in the theory of knowledge of the subject-object distinction. This is, at least, what I intend to prove in this section.

On the occasion of Harald Høffding's 85th birthday on 11th March 1928, at which time Bohr was finishing the revision of the Como manuscript, he made a public statement in the evening edition of the newspaper *Berlingske Tidende* of what Høffding meant to him and to other scientists of his age:

> On professor Høffding's 85th birthday the warmest feelings of gratitude will stream towards him from the younger generation for all that he has meant to us through his personal instruction and his publications. Standing, as I do, outside the circle of professional philosophers, I am not qualified to speak about his extensive and wide-ranging scholarly activity. If nevertheless I am prompted to make a contribution today by expressing our gratitude, it is, first and foremost, to the connection between philosophy and the natural sciences, which Høffding himself so often and vigorously has stressed, which I have in mind. This connection is one not only met at the very earliest stages of science, but is also a permanent one: cross-fertilization has taken place during the entire development of science up to recent times, where the vast accumulation of empirical data in all fields and the high standard of methods required for the acquisition and analysis of the data have necessitated extensive specialization in science.
>
> In the study of the phenomena of nature we are, time and again, faced with problems which call for a revision of the concepts underlying our understanding of observations. Every time there has been a crisis conditioned by external circumstances, originating in an apparent conflict between old and qualitatively new experiences, there has been a conflict foreboding an obstacle to the attempts of human thought to penetrate the secrets of nature. It has been of inestimable significance that scientists have been able to find support *and points of departure* for new advances in the endeavors of philosophers to make clear the foundation

and limits of human intellectual activity. Without being discouraged by the practical difficulties which face the scientist, the philosophers have, solely from a need for coherence and harmony in mind and thought, deepened our knowledge and have fostered a general attitude towards emerging difficulties and have furthered a widespread understanding of the relativity and *complementarity* of all human concepts. Because of his inevitably one-sided training the scientist is very often, in the particular, prevented from imagining to what extent the foundation, on which we are building, is created not only by the activity of pioneers whose names are connected with discoveries within the narrow domain of science, but also by the efforts made by great thinkers in dealing with concentration in problems which are common to all mankind; efforts from whose fruits we already benefit through the formulation of terms which, though expressing scientific ideas, have now become part of the common language. We are all very much indebted to Professor Høffding who, through his characteristically objective yet personal presentation of the conquests of philosophy in the struggle to elucidate the presuppositions of both the acquisition of knowledge and the life of the emotions, has contributed to the increase of our understanding of the foundations of our work, and has thus actively supported us in it.

Personally I have had the good fortune from my earliest youth to be in close contact with professor Høffding, and I feel that I owe him a great debt of gratitude for his instruction and encouragement. However, I would scarcely have seized the opportunity to express these feelings publicly, if it were not because I knew that I would at the same time be expressing sentiments of which a large circle of the younger generation of Danish scientists has a lively awareness. Høffding has not only guided us into the sublimities of philosophy, at once so remote and so close, but his unfailing freshness and openness of mind towards every new advance, the development of which, even the most recent, he has kept up with in every field, has if possible strengthened our confidence in him and won for him the affection of all. If it is the case, in this country, that hardly anybody looks upon philosophy as idle speculation about questions which are not capable of aiding mankind's quest for social progress and mastery of nature, but that we all praise philosophy as the science of sciences, then it is due first and foremost to the activity of Høffding and the tradition he has created. When abroad I too have had plenty of opportunity to experience how highly Professor Høffding is admired for the unique character of his approach to science in general. By colleagues he is recognized everywhere as a master with whom hardly any contemporary philosopher with regard to outlook and standards of objectivity can bear comparison. May he continue to be active among us with unremitting efforts for a long time to come.[5]

This is a consummate statement not only of Bohr's intellectual debt to Høffding, but also of Bohr's conception of the relation between science and philosophy, which was, after all, essentially the same as Høffding's.

As it happens, we know how Høffding reacted to Bohr's homage. On 23rd March 1928 he wrote to Meyerson:

Bohr declares that he has found in my books ideas which have helped scientists in the "understanding" of their work, and that they thereby have been of genuine use. To know this is a great satisfaction for me, who feels so often the deficiency of my own knowledge with respect to the natural sciences.

Høffding's impression of what Bohr meant corresponds very closely I think, to what Bohr sought to express.

Public remarks similar to those of Bohr's about Høffding were expressed by Edgar Rubin a few years later, shortly after Høffding's death:

Those of us, now advanced in years, who have worked in the field of philosophy and psychology were in close contact with Harald Høffding. He was our teacher, and he influenced us perhaps more than we were aware of. In cases where we developed independent opinions and positions, these came into being due to the fact that, to no negligible extent, we held our ground in the face of his view.[6]

And both Niels and Harald Bohr, Hatt, Brøndal and Rubin have also, on occasion, privately expressed a similar debt to Høffding as we have seen from letters in their hand now in the Royal Library.

From what has been argued up to now, Høffding's influence on Bohr was at least of a personal and intellectual nature. But we have also seen that Edgar Rubin, who first became a lecturer of philosophy and then a professor of psychology, has mentioned that he and other scientists of his generation were strongly influenced by Høffding as far as psychology and philosophy were concerned. Similar remarks are found in Bohr's obituary of Høffding presented in 1931 to the Royal Academy, where he also refers to one of Høffding's most important works *Psykologi i Omrids paa Grundlag af Erfaringen* (Outline of Psychology on the Basis of Experience), published for the first time in 1882.

I am grateful to have been asked to say a couple of words tonight, at a time when Harald Høffding's memory is being honored in the Royal Academy, which was always particularly close to his heart. Nobody would expect from me a penetrating account of Høffding's personal development and scientific endeavour, such as the one we have just listened to with awe; but taking my point of departure from those memories so dear to me which my own relationship to Høffding holds, I would like to attempt quite briefly to express what his personality and life's work has meant to large circles of those who have served science, and whose studies have been only indirectly connected with philosophy proper.

Bohr then gives the account of the gatherings in the home of his childhood which we already quoted on page 13. He continues by saying:

The accentuation of the unity of science of which our Society is a symbol was to Høffding not merely an abstract matter of fact but a practical necessity. Even though he would perhaps like to characterize philosophy as the science of sciences it was alien to him to believe that philosophy in a strict sense should establish the laws to which all scientific work should conform. Høffding was always prepared to accept that important aspects of the general human problem of knowledge can be viewed in a new way, benefitting from the studies carried out in the more specialized areas of science where particular features contribute to the general view of the mutual relationship between experiences. On the basis of this approach his chief endeavour was conceptually to underpin points of view developed within the different branches of science in order that they might shed light on general questions. The uniquely comprehensive view of the forms of scientific thinking so acquired enabled him in return to offer the scientist a lesson that was all that much more valuable because the ever extending ramifications of science make it still harder for the students of the various sciences to gain immediate understanding and learning from each other's work.

Even though the general problem of knowledge in the aforementioned sense was central to Høffding's concern and over the years increasingly so, it was, however, first of all psychological studies which enabled him to develop what was to become his characteristic method, and in which he fashioned his tools for the treatment of abstract questions. The framework which these studies provided for his work in other areas of philosophy is one to which Høffding has drawn attention in his last essay, a short paper entitled "Psychology and Autobiography", which he wrote only a year ago and which will be published in the near future in an American collection. In an extremely enlightening and interesting way Høffding here reviews his production and the basic approach behind it. It was an unforgettable experience, while the writing of the paper was in progress, to listen to him talking about his long life of work and to sense how he, through the recollection of the inner satisfaction it had given him, had found the strength to withstand the heavy sorrows which life had inflicted upon him, not least in later years.

The exposition of his theory of psychology which Høffding composed for his propaedeutic lectures and which he typically entitled "An Outline of Psychology on the Basis of Experience" was indeed also the work through which Høffding for the first time reached a larger and scientifically minded group of readers. The work, whose peculiar attraction and force of conviction were first of all due to the author's reverence for the greatness of the subject, had gained popularity and retained a vitality of which Høffding would scarcely have dreamt when he wrote it fifty years ago. One is not to look for poetic renderings of the movements of the life of the spirit or perspicuous explanations of normal and pathological mental states. On the contrary, one will find an account, at once sober and enthusiastic, of a scientific approach, in the truest sense of the word, to the life of the mind. The endeavour to retain a balance between analysis and synthesis is predominant; the fact that even though the whole consists of the parts, the appearance of the parts are influenced by the whole is never lost from sight. For many of those who heard his lectures, and for the even greater number who read his book, this characteristic objectivity in Høffding's presentation of psychology has certainly had a significance that is deeper than any one of us is easily able to express. This has struck me particularly when I've been with students at new universities, which have no tradition such as does an old college as ours, and have encountered a narrowing of the general scientific attitude which results from the lack of insight into the basic problems of psychology such as that which Høffding's pupils received quite spontaneously.[7]

At this point in Bohr's commemorative speech follows the other passage which we have quoted above on page 41, the one dealing with his discussion with Høffding about logic. In connection with the paper "Psychology and Autobiography" mentioned by Bohr it should be noted that Høffding was at that time regarded as one of the most outstanding psychologists in the world, and was elected, as the only psychologist in Scandinavia, to contribute with this paper to the series *A History of Psychology in Autobiography*.

Those who are familiar with Bohr's writings will know how often he compares the problems of observation in quantum mechanics with similar problems of observation in psychology. Sometimes he also mentions that the notion of there being complementary descriptions of various phenomena, which arises from a coherent solution to the problem of observation in quantum mechanics, had been recognized by philosophers much earlier in the field of psychology.[8] Recall, too, similar remarks by Bohr in the tribute he made to Høffding in the newspaper columns, where he wrote about those philosophers who have brought about a widespread understanding of the relativity and complementarity of all human concepts. Høffding was certainly the one whom Bohr primarily had in mind. In his book on psychology Høffding has in fact described these problems of observation in psychology, both in relation to self-observation and with respect to psychological experiments. And in his book on ethics Høffding discusses the complementary aspects involved in making a study of free will whilst engaged in the performance of an act. This is acknowledged by Bohr in the passage which follows that cited earlier:[9]

Høffding's exposition of the history of philosophy, which has won world-wide acclaim for its impartiality, for the patience with which he sought to examine the conditions under which the works of the great thinkers were conceived, and for the attempt to grasp the essence of their thought bears, more than anything else, witness to Høffding's general knowledge of philosophical systems and the conditions under which they came into being. This exposition acquires its individual character not least through the deep interest it shows for the evolution of the natural sciences and the understanding of their significance for philosophy in general which Høffding insists upon. In his later years Høffding's approach found a natural expression in his understanding of the revision of the physical conceptual framework brought about by the opening up of new areas of experience and whose relation to the theory of knowledge was the subject of his last lecture here in the Royal Academy. With a receptivity and freshness that was surprising in someone of his age, Høffding was fully sympathetic to the efforts of the physicists to extend the framework for ordering experience, and he was happy to recognize features in the new forms that he himself had encountered earlier and described in his psychological works, namely in connection with ethical questions. Indeed, many will perhaps – in the new light shed by the development of atomic theory on the problem of causation – first now be fully able to assess the perspicacity and the aptness of choice of expression that Høffding displayed in discussing the old riddle posed for thought by the freedom of the will.

The continuing development and clarification of Høffding's philosophical premisses which continued until death ended his long life was in the most intimate way linked to his individual method and entire way of thinking. Every time Høffding in this Society gave an account of his work, and was therefore provided with the opportunity to express his opinion on questions which he had examined at an earlier time, the attentive audience was invariably conscious of new aspects being inserted and of how his views were constantly being amplified, rounded out and brought into mutual harmony. To visit him in his last years was on every occasion a great and enriching experience. In spite of the sadness to which Høffding at times was subject, because of the anxiety wrought by the failing health of the one closest to him and the growing solitude caused by the demise of the friends of his youth, one always left him with a feeling of having been brought out of the commonplace and given fresh instruction on the depth and beauty of the harmony of existence. Undiminished was the love he retained for everything of worth he had learned to treasure. Towards the end of his life he talked with youthful enthusiasm about the poetry of life which he found in Plato and Spinoza as well as in Shakespeare and Goethe. When all is said and done it was this love and fidelity that made Høffding the true philosopher he was, whose death leaves so great a loss in many quarters.

As we shall see, in the last lecture he gave in the Royal Academy, which Bohr mentions here, it was clear that Høffding looked upon the concept of complementarity as one whose application he had earlier called to our attention in relation to psychological and ethical descriptions of one and the same action.[10] And we have just seen Bohr express the view that some of the features that Høffding had described in his psychological studies of free will had reappeared in new forms in quantum mechanics. So it is fair to conclude, I think, that Bohr's account of the problems of observation in psychology stemmed from Høffding's analyses of these problems; and that Bohr, in effect, acknowledged this himself.

There is much less evidence that Bohr recognized how similar his epistemology and ontology were to Høffding's. But there is strong evidence that Høffding did. One natural reason for Bohr's not having been struck by the similarities was that he had really only seen philosophy through Høffding's eyes. How similar their views were could be discerned only against the background of a range of epistemological and metaphysical views with which Bohr had but slight acquaintance but with which Høffding was very familiar. Bohr may also have looked upon philosophy as having a fair resemblance to physics in the sense that in philosophy genuine results had been amassed over the years in the same way as had been the case in physics, so what he had acquired from Høffding was familiarity with a very well-established tradition in philosophy.

Regardless of what it was that brought it to his attention, Høffding did notice the similarity when he wrote in his notebooks a short piece of five folio pages, entitled *Nogle Bemærkninger om Årsagsprincippet og den moderne Elektronteori* (Some remarks on the principle of causation and the modern theory of the electron), but the pages were later removed. It must have been written around the spring of 1928.[11] It was probably Høffding himself who was responsible for cutting the missing pages out of his notebooks in order to send them to Bohr, as in a letter written to Bohr in the summer, now in the Bohr Archive, he refers to a paper of this sort:[12]

Carlsberg d.11 July 1928

Dear Professor Bohr,

My wife and I were very sorry that we had to abandon the plan to visit you. My wife is now at Dr. Borgbjærg's gastric clinic for an ailment which has troubled her for a long time and she sends her thanks for the kind letter from your wife. She feels well at the clinic and has started to devote herself to literature again. Perhaps then the cause of her ill-health has been found.

I am still engrossed in your last essay,[13] and even though I am reluctant to disturb you during your vacation, I cannot refrain from asking you if the enclosed draft shows whether I have correctly understood the reasoning in your latest works in as far as they are concerned with an epistemological problem. I am in no hurry for the answer, and I hope in any case that after the vacation we may have a talk about the relevant topics.

It is, indeed, chiefly in psychology that the question of the possibility of the principle of causality is raised. All the more interesting is it, then, that it is now being raised in physics.

With kind regards and thanks to you for everything.

Yours sincerely,
Harald Høffding

We can already see from this letter, which must have accompanied Høffding's essay, that Høffding was the first explicitly to regard the difficulties of applying the principle of causation in quantum mechanics as analogous to the similar problems of using the principle of causation in psychology. Høffding also indicates here that the principle of causation has invariably presented difficulties in psychology, and that in the light of this, the new situation in physics is most interesting.

Only twenty days after Høffding had sent his paper to Bohr, he received a letter from him in which the essay is spoken of in very positive terms.[14]

Lysthuset Tibirkelunde, Tisvildeleje 1–8–1928

Dear Professor Høffding,

I am very sorry about not having written until now, but I got your letter with the interesting and thought-provoking essay on your views concerning the principle of causation just as I was about to go on a sailing trip to the Swedish and Norwegian skerries with Bjerrum and Chievitz. I believe that I need not tell you how much it pleases me that you think that you could perhaps make use of the as yet very unpolished remarks with which I have tried to state the grounds, with respect to the analysis of the phenomena of nature, to which the development of the quantum theory led the physicists to endorse.

In so far as I am qualified to follow your train of thought, I believe that I can accede wholeheartedly to your opinion regarding the thoughts that have dominated work in the area of atomic theory in recent years. It is indeed especially the purely epistemic side of the analysis of the concepts that I have had in mind in my work and with which the final remarks in my paper are concerned. Lately I have been working on a further analysis of the concept of observation as it is used in the presentation of the physical forms of perception, and I hope to be able to present the question of the foundations of the description of nature a bit more clearly than I did in my essay, even though in this respect I am more than ever acutely aware of my lack of philosophical knowledge. As far as the psychological problems referred to in your essay are concerned, I feel this gap in my knowledge even more keenly, were it possible. Yet I have been powerfully struck by the possible scope of the general considerations with which you conclude your essay. Sometimes I have the vague idea that there might be a possibility of proving a similar complementary relation between those aspects of the description of the individual psychological processes which relate to the emotions and those that relate to the will as that which quantum theory has shown to obtain, with respect to elementary processes in physics, between the conservation of momentum and energy on one side and the space-time coordinates on the other. Yet the difficulty of establishing such an analogy in every respect may first and foremost be the result of the impossibility in the field of psychology, given at least its present stage of development, of putting forward a definition of an elementary process that possesses the simplicity and determinacy similar to that which can be obtained in the field of physics with the aid of the quantum postulate.

Whilst on the trip I made the decision many times to write more explicitly to you and thank you for your letter, but life on board has never given me the leisure to do so. This is precisely what is refreshing about a sailing trip: the very conditions of one's existence changes, so to speak, from moment to moment in accordance with the unpredictability of the weather and the sea. We had a pleasant trip and everyone on board asked me to send you and your wife their kind regards. I postponed writing to you until I was back at home, since I had also hoped by that time to be able to make an arrangement with you about when we could expect the visit from you both, which my wife and I had looked forward to so much and which would provide me with the best opportunity of learning more about your views. However, when I first arrived here yesterday I heard about young Harald's illness, which has given rise to much anxiety. Unfortunately, as my wife has written in her letter to yours, we must under these circumstances forgo the pleasure of seeing you here for the time being. However, as soon as I come to the city I will pay you a visit, and I hope then to hear and to bring good news about everybody's health.

With many kind regards from both of us and from Mother too, who had looked forward to coming here at the same time as you and your wife.

Yours sincerely,
Niels Bohr

Bohr seems to have been very anxious to discuss Høffding's paper, for two weeks later in his letter of 13th August to Meyerson Høffding writes, "I have had most interesting talks with Mr. Niels Bohr especially on the irrationality brought into physics by the theory of quantum mechanics". Unfortunately, we do not know where Høffding's paper is now. It would seem that it has been lost. But Høffding's and Bohr's letters tell us, at least, that as early as the spring of 1928, about the time Bohr's Como lecture "The Quantum Postulate and the Recent Development of Atomic Theory" from 1927 was published in *Nature*, Høffding must have outlined some general parallels between the complementary mode of description in quantum mechanics and psychology, a fact which Bohr first seems to recognize in his two papers from 1929, "The Quantum of Action and the Description of Nature" and "The Atomic Theory and the Fundamental Principles Underlying the Description of Nature". We also learn of Bohr's feeling that his knowledge of philosophy, and of psychology in particular, was deficient. However, two years later Høffding published another paper, the last he was to publish, on epistemology, which probably contains elements from the earlier one. This paper will be examined in the next section.

Bohr's Como lecture, in which he presented the notion of complementarity for the first time, was delivered at Como in Italy on 16th September 1927. One month later Bohr gave the same address at the Solvay Conference in Brussels, but owing to an essentially negative response Bohr felt forced, during the winter of 1927–1928 (partially assisted by Wolfgang Pauli), into successive rewritings of the original talk. The result was that the paper known as "The Quantum Postulate and the Recent Development of Atomic Theory" appeared in one version in the Congress Proceedings and in a substantially revised version in *Nature*. It is also the latter which Bohr included together with the two papers of 1929 in his collection of essays entitled *Atomic Theory and the Description of Nature*, which was published in 1931 in German and in 1934 in English.

Thus the classical paper "The Quantum Postulate and the Recent Development of Atomic Theory" was published for the first time in *Nature* on 14th April 1928. However, Høffding referred to it as early as 30th March 1928 in a letter to Meyerson, so apparently he must have read or discussed Bohr's manuscript before it was printed. This inference is reinforced by the fact that no excerpt from this essay exists among the many extracts which Høffding had transcribed of every, or nearly every, book and paper he ever read. We know that Høffding must have been familiar with Bohr's ideas at least from a session in the Royal Academy on 18th November 1927, when Bohr gave a talk in Danish on "The Quantum Postulate and the Recent Development of Atomic Theory", Høffding being among the audience.[15]

It is first in his papers from 1929 that Bohr focuses on the problem of observation in psychology. But, the classical paper from 1927 ends with the following "hint": "I hope, however, that the idea of complementarity is suited to characterize the situation, which bears a deep-going analogy to the general difficulty in the formation of human ideas, inherent in the distinction between subject and object".[16] Since Høffding's letter to Meyerson bears evidence to the fact that he must have read an earlier draft of Bohr's *Nature* paper, it is very tempting to infer that Bohr's last sentence reflects something of what Høffding had pointed out to him in discussion before the paper was printed. This suggestion is supported by the fact that Høffding, commenting on Bohr's paper to Meyerson, wrote: "Here he tries to overcome the difficulty which lies in the fact that the electron has simultaneously to be a particle which is located at a definite position and to be a source of energy. Here we have *an old problem presenting itself once more at the frontiers of the natural sciences*" (my emphasis). This relates directly to the last sentence of Bohr's paper. Furthermore, the suggestion is also confirmed by the fact that the paper which appeared in *Nature*, and which is normally called Bohr's Como lecture, is not the one which he delivered at Como on 24th September.

If the two versions of Bohr's Como paper are compared, we find that the last statement in the *Nature* version, the one quoted above, was not in the original Como version. So it must have been added to the version to appear in *Nature* between November 1927 and March 1928. It is also the only sentence in the final version of "The Quantum Postulate and the Recent Development of Atomic Theory" in which Bohr refers to an analogy between the observational situation in quantum mechanics and the difficulties of making a subject-object distinction in the human sciences and, apparently, nowhere in his many brief written remarks for the preparation of the original talk does he mention this distinction. So in the light of the above considerations it is surely very compelling to draw the following conclusion: that at some time between November 1927 and March 1928 someone must have pointed this analogy out to Bohr; and who other than Høffding?

Further evidence for this conclusion may be found in Bohr's use of the term "irrationality". In one place in the original talk given in Como Bohr characterizes Planck's discovery of the quantization of energy as "the irrational element expressed by the quantum postulate". However, the sentence just preceding the last statement on subject and object in Bohr's *Nature* version is the following: "In the quantum theory we meet this difficulty [that our language refers to our ordinary perception] at once in the question of the inevitability of the feature of the irrationality characterizing the quantum postulate". Here Bohr recapitulates his fundamental premise in that paper that the quantum postulate, which presents the problem of observation in quantum mechanics and, properly interpreted he claims, leads to the relationship of complementarity, represents an "irrational" feature. Such a characterization is also made at an earlier point in the paper, where Bohr uses the phrase "the quantum postulate with its inherent 'irrationality'". These two statements were also first put into the essay between its presentation in Como and its publication in *Nature*. But, as we shall see later on, there was every good reason for Bohr to emphasize this point. Høffding had, indeed, characterized the distinction between subject and object as one of three "irrational" elements in cognition. It must have been Høffding who recognized that Bohr's talk about "the irrational element expressed by the quantum postulate" and the difficulties confronting an objective description of quantum phenomena as a consequence of the quantum postulate were part of the general problem of distinguishing between subject and object in human thought.

The subject-object distinction too surfaces for the first time in Bohr's scientific correspondence of 1928. In a letter on 24th March to Dirac, who had just been involved in the work on the proofs of the Como paper to *Nature*, Bohr wrote with a reference to "the endeavour [in my article] to represent the statistical quantum theoretical description as a natural generalization of the ordinary causal description":

> In this respect it appears to me that the emphasis on the subjective character of the idea of observation is essential. Indeed I believe that the contrast between this idea and the classical idea of isolated objects is decisive for the limitation which characterizes the use of all classical concepts in the quantum theory.[17]

Bohr seems, in the course of 1928, to have considered the problem from every angle in discussions with Høffding. Undoubtedly, influenced by these discussions and by the content of the paper from Høffding's hand which has not survived, Bohr started, in the autumn, on the preparation of his contribution to the celebration of Planck's 50th doctoral anniversary, entitled "Wirkungsquantum und Naturbeschreibung" ("The Quantum of Action and the Description of Nature"), which was published in a special jubilee issue of *Naturwissenschaften* on 28th June 1929. On this occasion he wrote on 7th November in a letter to his Swedish friend Carl W. Oseen, the physicist:

> My article in Naturwissenschaften is of course concerned with a general attitude, which I have had at heart during all the years that I have been occupied with the quantum theory, but which we have only got the means to express through the great development of recent years, which has made possible a consistent representation of the experimental evidence. As we already discussed years ago, the difficulty in all philosophy is the circumstance that the functioning of our consciousness presupposes a requirement as regards the objectivity of the content, while on the other hand the idea of the subject, of our own ego, forms a part of the content of our consciousness. This is exactly the kind of difficulties of which we have got such a clear example in the character of the description of nature required by the essence of the quantum postulate. Far from bemoaning the fact that in atomic physics our usual wishes with respect to the description of nature cannot be fulfilled, I believe that we ought to rejoice at the new lesson concerning the limitation in the human forms of visualization that is implied by the discovery of the quantum of action.[18]

Although Bohr mentions here that the problem of the objective content of a subjective consciousness had exercised his mind for as long as he had been occupied with quantum physics, it does not follow that Høffding had no part in it when for the first time, in the jubilee paper, Bohr extended the idea of complementarity to psychology and epistemology in general. On the contrary, it tells us that Bohr had been influenced very early on by Høffding with respect to the centrality of the subject-object problem in epistemology. But in spite of this one may doubt whether Bohr had any clear or substantial idea of what "the general lesson of quantum mechanics" was, until Høffding pointed out a general epistemological similarity between psychology and quantum mechanics.

An anecdote may partly confirm Høffding's early influence on the interest Bohr took in the subject-object problem in epistemology, even though the exact reproduction of the episode is rather dubious. Bohr's latest assistant Jørgen Kalckar reports the episode – which he believes illustrates Bohr's very early anticipation of the epistemological lesson of quantum mechanics – by saying that when Bohr once lectured Edgar Rubin in the late twenties on the "lesson" of quantum mechanics, the latter was to have exclaimed: "But Niels! You told us all that twenty years ago".[19] Nevertheless, it is doubtful, I think, whether it was Rubin – with his intimate knowledge of Høffding's philosophy – who made that remark. According to Heisenberg, who actually heard the utterance, it was made by one of Bohr's friends on a trip from Copenhagen to Svendborg in Funen in Bohr's sailing boat *Chita*, but he does not say by whom. Yet, Heisenberg mentions two of the friends on board by name, one of Bohr's colleagues, the chemist Niels Bjerrum (1879–1958), and the surgeon Ole Chievitz (1883–1946).[20] Further, from the context there seems to have been at least one more friend on the sailing trip. Most probably this was the artist Holger Hendriksen (1878–1955), who, in a syndicate with Bjerrum and Bohr, owned the boat and to whom Heisenberg refers indirectly. This makes five persons in all. Heisenberg does not refer to any other person; thus, it is hard to believe that Rubin should have been a sixth member of the party on board. Moreover, in Heisenberg's rendering of the remark the number of years mentioned is not twenty but only ten.

By the autumn of 1928 Bohr's was not the only mind to be engaged by the problem of the objectivity of descriptions of experience in quantum mechanics and psychology. As mentioned above Høffding wrote another essay on the latest developments in the theory of knowledge which was published in 1930. From Høffding's Notebooks it may be established that he made four successive drafts for *Bemærkninger om Erkendelsesteoriens nuværende Stilling* (Notes on the Present State of the Theory of Knowledge) before being satisfied with it. It seems as if the first one was written during the autumn of 1928 or the winter of 1928–1929.[21] We know that Niels Bohr discussed that paper with Høffding too and probably read it before it was published, because Høffding first presented it as a lecture in the Royal Academy, on 17th January 1930, and prior to doing so he sent the following letter to Bohr.[22]

Carlsberg, 4th Dec. 29

Dear Professor Bohr,

We did not make any arrangement the other night. But I would appreciate it if I could present you with a rough draft of a talk I am thinking about submitting to the Royal Academy just after the New Year. If it is possible for you to give me a few hours of your time, at the end of this week or at the beginning of the next, I should be very pleased.

I was glad to see you at the meeting of the Philosophical Society. I would like to have listened to what Jørgensen had to say, but I was too tired.

With kind regards,
Yours sincerely Harald Høffding

Present on the evening when Høffding gave his speech in the Royal Academy were, besides Niels Bohr, three other former members of *Ekliptika*, namely Harald Bohr and Poul and N.E. Nørlund.[23]

The meeting of the Society of Philosophy and Psychology to which Høffding refers in his letter to Bohr took place on 28th November. First the Society held a general meeting, and when that part of the session had ended, the invited speaker, Niels Bohr, gave a talk entitled "Nogle Bemærkninger om den nyere Fysiks stilling til Aarsagssætningen" ("Some remarks on the relation of the new physics to the principle of causation").[24] After Bohr's talk there was a discussion in which Høffding, Hatt, whom we recall from the Ekliptika Circle, and Jørgen Jørgensen participated. Since the meeting dragged on, finishing at 12.30

a.m., Høffding retired before Jørgensen's contribution. A few days later, on 7th December 1929, Bohr wrote to his friend and former assistant, Hendrik Kramers:

> By the way, I gave a lecture about this [the problem of causality] to a Copenhagen association, which calls itself the Society for Philosophy and Psychology, and I learned a great deal from the ensuing discussion. In particular, I know better which points non-physicists resent, and I also believe that for this very reason I found on this occasion better words than previously to answer the objections.[25]

It is apparent that at that meeting Bohr was for the first time confronted with serious criticism from non-physicists – especially from Jørgensen, who was never elected as a member of the Royal Academy – objections which Høffding had not contemplated. Two months later he was publicly to express his approval of Bohr's ideas of complementarity in the Royal Academy.

There is, nevertheless, a fly in the ointment. Present at the meeting of the Society of Philosophy and Psychology was H. Fuglsang-Damgaard (1890–1979), who was later to become Bishop of Copenhagen. This emerges from a correspondence, now in the Bohr Archive and to which David Favrholdt has drawn my attention, between the Bishop and Bohr on the occasion of the centenary of Høffding's birth, a correspondence which seems to contradict what has gone before. In a letter, dated 18th January 1943, Fuglsang-Damgaard writes:[26]

> Dear Professor,
>
> On the occasion of the centenary of Professor Høffding's birth I intend to write an article about him in "Teologisk Tidsskrift" (Journal for Theology).
>
> I am sure you will recall giving a lecture, about 12 years ago, at a meeting of the Society for Philosophy and Psychology at which Professor Høffding was present and took the floor. He said something to the effect that he, having for two years studied your thought, had come to the conclusion that the foundation of his philosophy was no longer valid.
>
> His words made on me an indelible and unforgettable impression. However I should be gratified if you were able to confirm this recollection. Thanking you and sending my kindest regards to you and your wife.
>
> Yours sincerely,
> H. Fuglsang Damgaard

Two days later, 20th January 1943, Bohr answered the letter:[27]

Dear Bishop Fuglsang Damgaard,

Thank you for your cordial letter. I remember Høffding's participation in the discussion following upon my lecture at the Society for Philosophy and Psychology very well but I do not recall the exact words which you quote in your letter nor the connection in which they were uttered. In particular, I do not remember whether they were uttered as a direct comment upon the lecture or as a response to a question put by someone else who was present. Høffding's whole attitude was to a singular extent marked by his openness towards what might constitute progress, also in fields where he himself had done a lot of thinking, and I imagine that his statement was meant as much as a warning against all forms of prejudice as a declaration of the relation between what had been the work of his lifetime and the problems posed by the new physics for the theory of knowledge. Shortly before his death, however, Høffding gave, in an article in the philosophical reports of the Academy of Sciences and Letters, "Bemærkninger om Erkendelsesteoriens nuværende Stilling" (Notes on the Present State of the Theory of Knowledge), sublime expression to his view concerning these problems and I believe it would be best, in order to avoid any possibility of misunderstanding, if you were to find a statement in this article suitable for your purpose. Hoping that it will interest you I enclose copies of some articles which I have published over the years in "Naturens Verden" (The World of Nature) which concern the general problems of the theory of knowledge touched on above.

Yours sincerely,
Niels Bohr

In his reply Bohr does not directly deny the authenticity of the Bishop's recollection but neither does he confirm it. From what he says in this letter it seems most likely that he cannot recall anything like the words which Fuglsang-Damgaard ascribes to Høffding. And if the remark in question had reflected Høffding's considered opinion Bohr would have been acquainted with it from their many discussions in private and would thus have been able to confirm it on the basis of those. Moreover, if Høffding had felt that the foundations of his philosophy were opposed to Bohr's interpretation of quantum mechanics, he would, indeed, have objected to it much earlier, and Bohr would therefore have known which points "non-physicists resent" before the meeting at the Society for Philosophy and Psychology. Being a singularly courteous and deferential person Bohr would not directly contest Fuglsang-Damgaard's memory or indicate that the Bishop was mistaken, especially since Høffding's words had made "an indelible and unforgettable impression" on the latter. Instead Bohr expresses himself cautiously and tentatively. He tactfully points out that if Høffding had said that he "had come to the conclusion that the foundation of his philosophy was no longer valid" one has to consider in which context the statement was uttered. I take Bohr to be indirectly expressing his doubt as to whether Høffding had made such a sweeping statement. But, since the Bishop's recollection is not in accord with his own of the evening in question, and his overall understanding of Høffding's philosophy, he advises the Bishop to look into Høffding's "Bemærkninger om Erkendelsesteoriens nuværende stilling" "in order to avoid any possibility of misunderstanding" and to see whether he could find some confirmation of what formed the gist of his impression. Bohr was well acquainted with the paper, he had discussed it with Høffding prior to its publication, and, as we also have seen, returned to it in the last interview he gave, characterizing it as the best essay on complementarity written by a philosopher. So he assumed, I surmise, that there was little in it to support the Bishop's recollection. And, indeed, he was right. We shall see that it is more correct to say, as Aage Petersen remarked in the last interview, that Høffding in this paper "wrote mainly about his own anticipations of" complementarity. It would therefore be strange indeed if Høffding had written a paper on this theme in the period immediately prior to the meeting and there to declare that Bohr's thought rendered his own philosophy invalid.

What it was Fuglsang-Damgaard may have heard Høffding saying might, I suggest, have corresponded roughly to the following. He, Høffding, had always believed that the application of the concept of continuity is fundamental to any rational understanding of nature, but quantum mechanics confronts us with discontinuous processes. So the existence of such elements may seem to invalidate his philosophy in the sense that probabilistic features are introduced into the description of atomic objects in a sense which is at variance with what constitutes a continuous description. However, Høffding had always regarded as admissible the existence of discontinuities which would present themselves as irrational elements for our cognition in virtue of marking the limits of the application of the forms of thought and perception. So Bohr's interpretation of quantum mechanics did not represent a serious challenge to Høffding's philosophy, and apparently this was also recognized by Fuglsang-Damgaard after having read Bohr's letter and Høffding's essay. In his centenary paper on Høffding the Bishop refers to the episode after having summarized part of the paper. "Towards the end of his life Professor Høffding, at a meeting at the Society of Philosophy and Psychology where a talk had been given by Professor Bohr, made some observation upon the consequences of the new physics for his own philosophy. The aged scholar's integrity, and the energy with which he approached new problems, made a deep and unforgettable impression".[28] The Bishop's recollection, then, was no longer that Høffding had claimed that the foundation of his own philosophy was invalid, but only that the philosopher, then advanced in years, acknowledged that the new situation in physics might imply further progress in the theory of knowledge and modifications in his own philosophy.

During the one and a half years prior to his death Høffding produced almost nothing. Shortly after his death there was once again an opportunity for Bohr to disclose the nature of his latest meetings with Høffding. In August 1932 the

Tenth International Psychology Congress was held in Copenhagen – of which Rubin had asked Høffding two years earlier on the behalf of the organizing committee to be the president – and all the participants were invited by Bohr to the honorary residence at Carlsberg, which he had taken over after Høffding. In addressing a welcome to the guests he spoke in English about Høffding's approach to physics and its relationship to psychology. Unfortunately the lecture has only been preserved in a carbon copy in which several words and sentences are missing, and from the fact that the sentences lack polish and the English is faulty there is room for doubt as to whether the surviving copy is the final version. In spite of this it still gives us some information about the discussions he had with Høffding. So here is what Bohr said, or something very like it.[29]

It is to my wife and myself a very great pleasure and honor to welcome so many distinguished psychologists here, where Harald Høffding spent the last 20 years of his long life, and where he found opportunity to complete several of his great works. I have had the privilege of being in close contact with Høffding from my early youth as my father was an intimate friend of him, and I have at all stages of my life been able to benefit from the true scientific and philosophical spirit which you know from Høffding's work, but which found special expressions by personal acquaintance. I have not the qualifications to give proper judgment of what Høffding has developed, and I will just say that it was a very great experience to come and visit Høffding here in his last years. In spite of physical weakness, his mind was always active, he was always endeavouring on rounding and revision of his views to be able to take up any new knowledge, which life brought him and especially to take up the proper attitude to give the continuous development of the various branches of science. Before I said that I have not the qualifications to give the proper judgment of Høffding's achievements, qualifications which so many of you possess to an eminent degree. Neither have I the gifts of reproducing the discussions I had the benefit to have with Høffding in a proper dramatical form. I shall therefore just try in a few plain words to tell what was the questions under discussion and try to give you an impression of Høffding's attitude towards life. Now the thing was the difference between physical science and psychology, one might think that such relations must be of a very distant character. One might perhaps define physics as that which remains of our description of natural phenomena, when all those aspects which have especial interest for psychologists are eliminated, but of course no such [lacuna] in physics need not say how nature works, but it was what we are able to say about it, what views we are able to say about it, what views we are able to communicate to each other and in that sense the elimination of the psychological aspects is in itself a psychological problem, and one which has some line especially instructive because the problems with which one is dealing in physics is so much simpler than many problems in proper psychology, and that therefore certain aspects of philosophy appear more pure and forceful than [lacuna]. Now the interest for physical science from such a point of view was one which Høffding had through all his work. It appears in all his writings and his form of psychology he constantly refers to physical laws not only as a background for discussion of the situation of living beings, which in certain sense again is the background of psychology, but also as a mean of developing and purifying philosophical views themselves. In this last point of view it appears perhaps most strikingly in Høffding's account, so very well known of newer philosophy, since the [lacuna], where he shows his intimate knowledge and interest for the work of Galilei and Hume, who at that time laid the foundation of modern physical science. Høffding very often emphasized, how such fundamental ideas as causality, how much the development of purification of such an idea is due to Newton's treatment of the solar system. We have very simple cases of a causal [lacuna] if we know the proceedings of the philosophy and psychology. [Lacuna] able to predict the position and velocities at some later moment, but how simple and how fundamental such views may be, they will in a development of science most rapidly be taken up to revision, and in our times we have the experiments of fundamental revision of the ideas of mechanics. I think first of the step known as the theory of relativity; in Newton's description it was an especial simplification that all forces were considered as working instantly, but through the development of science [lacuna] of all forces were. Einstein led to fundamental revision of the space-time coordination of physical phenomena, it became clear that concepts so fundamental as scientific ingenuity was more relative than I think even any philosopher had dreamt of. I do not mean to press that [lacuna] of the physicists over the philosophers. [Lacuna] is so very much simpler, and the situation becomes perfectly clear, that previous ideas are insufficient. Now the attitude Høffding took to the theory of relativity is characteristic for his whole scientific spirit. It was of course tempting for philosophers to criticize the more or less hard way in which the common physicists used or misused concepts and notions developed by the ingenuity of philosophy, but Høffding was far from such a [lacuna]. He took the new discoveries as well as regards the physical phenomena as setting starting point, {deletion: as regards the pure form} and tried if in his psychological experience he could find or recognize analogies between similar situations in order to make the new progress fruitful in psychology in the same time for himself to create a proper background for his attitude towards the physical points, but in these last years it was not the theory of relativity or relativity problems which was the scheme of our discussions, but it was the next step in the revision of the mechanical ideas originated by the great discovery of Planck of the quantum of action. Now I shall not try to give an idea of what quantum of action means. It is impossible, because the existence of the quantum of action is the feature quite foreign from ordinary mechanical ideas. We may even say that the existence of the quantum of action is a direct [lacuna]. Now therefore to look at this discovery became first gradually clear through its application, through its consequence at various branches of physics, but it found in the atomic theory [lacuna] and to avoid misunderstanding I should just like to say that the atomic theory has always attracted the interest of philosophers, I might even say it has been created by philosophers as Democrit. The division of science in later times was not a serious division, a division which on one hand is the background of modern science, and is the only way to avoid dilettantism. On the other hand it has attractive effects, if it were not for such men and spirits as Høffding, who tried to help in the common understanding by looking at the general background for what we at the various moments understand by an explanation: Now I wanted to say that the {deletion: whole} old philosophers interest in atomic theory on one side it is necessary for understanding the stability of natural phenomena [lacuna]. On the other hand one thought that this idea could never be brought to a real [lacuna] to prove the existence of individual atoms as sense observations involved. [Lacuna] number of atoms, but we all know now that this skepticism went too far from the development. The art of experiment has sharpened our sense to a degree that we are able to say that single atoms to recognize effects of ultimate particles of which atoms are built up; but we have in the same time been very forcibly reminded that we are here on new ground. We are outside the region of common daily life experience. Even if we may know the number, mass and particles of which the atoms are built up, we cannot account for the behavior of ordinary mechanics. This is due to the possibility of variations, which is quite opposite to the characteristic properties of the elements of the stability of natural phenomena. We are forced to assume that the state of an atom can only be changed by steps that any process which follows exchange of energy will be an individual process, by which the state of the atom is changed from one of its so-called stationary states to another of these states, and we can by physical experiments measure the possible energy values of an atom by the study of exchange of energy by collisions between atoms; but these transition processes between the states cannot be further analyzed by the concepts of ordinary mechanics, and it is also clear that the situation is different from the ordinary mechanical description. If we try to get to know something about the position of a particle in an atom, just as we measure the position of the moon, then we must necessarily use some tool of observations, some kind of measurements and now just due to the existence of the quantum of action. The interaction between atoms in such instruments will be of a similar individual character, and the use of the instruments means, that we allow an exchange of energy between the atom and the instruments which completely hide the energy balance, which we are allowed to investigate. We see that a description of the energy change of the atoms and a description where we are bound to study the position of the atom, are exclusive to one another and represent what is called complementary aspects in this way that they represent aspects of phenomena which exclude one another, but which are both necessary for a full description. Now this may sound very difficult, but it is something very simple which at the moment by physicists appear quite clear. If we study various aspects of such behavior of an atom, we must use various experimental instruments, various measuring instruments, and thereby we have to do with quite a definite phenomenon, when we investigate the various aspects. The common description which is independent of observation is an idealization which can be used for the motion of ordinary bodies. [Lacuna] which is very large compared with the quantum of action, and therefore we can do so without essentially disturbing interference with it. Now this is the situation which is completely new in physical science, and which all physicists will have to work very hard to get accustomed to, but that a man at the age of Høffding in his last years was prepared to take up the serious discussions to which the endeavors of physicists lead, is remarkable, and he was especially interested in the remarkable analogies with psychological phenomena. In this situation which is so new in physics is of course not new in psychology. In psychology the main difficulty is that of analyzing with interference, if we try to study some aspects of mental situations and mental processes, we have to direct our attention to it, and is only too well known to psychologists, and the situation is essentially changed, and in psychology we find a number of examples for such complementarity. An account of Høffding's attitude to such problems he has given in his very last paper, where he especially gives expression for the pleasure in recognizing in this new situation features which he had various times studied in his psychological researches, and I may say at the same time that remarkable for the [lacuna] of the spirit is the caution by which he expresses himself by the belief we have here a field where psychologists and physicists may be of great mutual help. The psychologists in offering all their studies of much more complicated situations which offers a background for the physical progress, and the physicists in offering the psychologists a special simple example which can be studied in great detail. Especial simple examples of the situation of the difficulties, with which they have to strain, [lacuna] but here form both sides we have to exert a great caution, because it is a great difficulty for people, who have not worked themselves in the field of science, properly to appreciate the strength and the deeper sense of arguments, and I think that in contrast to Høffding such caution is not always exerted neither by physicists nor by psychologists thinking about these [lacuna]. For instance I may just say that a problem which we have discussed in recent years is the question of the possible new aspect of physical development, all discussions of the freedom of the will [lacuna] that this background of classical mechanical ideas opens new possibilities for the behavior of the spectral influence, where we have to do with individual processes in our description reduced to a consideration of the probability of their occurrence sometimes in one or two steps, and we can speak of the probability taking one or another cause. The mechanical description may leave [lacuna] probability of individual processes in some way or other. To my mind such utterances are very dangerous and are very difficult properly to define. I may even say to a certain extent [lacuna]. The use of statistical methods does not mean that we in certain places stop with mechanical description, but is entirely bound by the formulation of physics, as the laws of atomic stability foreign to mechanics, and I think that as regards such problems as freedom of the will we cannot say anything else, that we here deal with forms of consistent life, the parallel of which in physical nature in the sense of Spinoza are not open to analysis by mechanical ideas. Such utterances on which I shall not enter further may appear very mystic, but it is the [lacuna] just to follow further out the lesson we are reminded of in physics of how much sense of the claims we can put to an explanation continuously changed with the development of science, and I can say that this lesson was perhaps that which was foremost in the mind of Høffding to give in all situations. With what I have said here it has been my intention to the best of my power re-create the atmosphere which Høffding passed to his surroundings, and I ask your forbearance if in what I have said I have myself not used the caution which I mentioned. I am sure I have used many words in a way, which does not quite correspond to the way in which these words are used by proper psychologists, but I hope that you will just have taken them in the same spirit as Høffding used, which is seriously in the mind of a physicist.

First, Bohr informs us that throughout his life he had opportunities to benefit from Høffding's mind and philosophical training through personal contact. He then proceeds to relate something of the many visits he had made to Høffding's home where they discussed the latest developments in atomic physics. Indeed, pivotal to their discussions were the epistemic differences and similarities between psychology and atomic physics. Bohr points out that already years before these discussions took place Høffding had shown interest in physics by including chapters on Galileo, Hume and Newton in his book *Den nyere Filosofis Historie* (A History of Modern Philosophy). Earlier on Høffding had likewise demonstrated his open-mindedness with respect to revolutions in physics by not criticizing the theory of relativity on the basis of certain philosophical preconceptions. Instead he accepted the recent advances in physics and made it his endeavour to find analogical features in the sphere of psychology with an eye to using such analogies to create a broader context leading to further understanding of those advances. Later in his talk Bohr indicates that in their latest discussions it had become clear to them that there is a remarkable analogy between psychological and atomic phenomena in the sense that the observational results of both kinds of phenomena derive in part from the interference that obtains between the state under investigation and the means of observation. Bohr also mentions Høffding's delight at recognizing the tenability of this analogy and being able to write about it in his last paper, "Bemærkninger om Erkendelsesteoriens nuværende Stilling" (Notes on the Present State of the Theory of Knowledge), convinced that the new epistemological situation in quantum mechanics bore a resemblance to that which he had described earlier in psychology. At the end of his address Bohr mentions their having discussed the problem of the freedom of the will in relation to the new development in physics, and in spite of many lacunae in the text and the obscurity of the formulation the proposed solution that emerged from their discussions seems to have been one based on a distinction between mind and matter considered as parallel to each other "in the sense of Spinoza". At the same time Bohr issues a direct warning against believing that an explanation of free will can be given in terms of indeterministic physical processes.

So far as it has been possible to reconstruct the matter of the discussions between Bohr and Høffding, it does seem right to argue that Høffding himself looked upon this aspect of the latest development of quantum mechanics as a counterpart to his own psychological and epistemological studies. We have also seen that to a certain extent Bohr has granted that this was the case, particularly with respect to his psychological studies. This claim may find further support in a letter from Bohr to a Finnish philosopher, Dr. Kalle Sorainen, who had written to Bohr on 12th July 1946 to ask him some questions about Høffding. Sorainen was staying in Copenhagen and working on the development of Høffding's epistemology. He conjectured that Bohr's theories had a certain influence on Høffding's thinking in the last phase of his working life. A month later Bohr replied to Sorainen from his summer cottage, where he was staying.[30]

10th August, 46

Dear Dr. Sorainen,

I am very sorry that, owing to a trip to England in July, I have not until now been able to answer your very nice letter of 12th July concerning your studies of Høffding's theory of knowledge.

From my earliest youth, owing, in particular, to the close friendship between Høffding and my father, I had plenty of opportunity to listen to and talk to Høffding for whose general humanistic and scientific values, I grew to feel an increasing sympathy and admiration. As can been seen from Høffding's latest work I had, in the years before his death many deep and searching discussions with Høffding on recent developments in physics. However, we went into no details about experimental results or their mathematical formulation, but merely into the general instruction with which the developments had provided us, which were, at that time, quite unfamiliar to many philosophers, but of which Høffding showed an exceptional understanding. The remarks in Høffding's last work on epistemology are based to a large extent on the study he had made of some papers I had written during those years and in which I had tried to make up my mind about some of the more fundamental questions, both so as to provide a contribution to the discussion going on among physicists and also as an attempt to further understanding in wider circles, without the specific use of mathematical methods. In the light of this, isolated statements in Høffding's essay cannot be taken too literally [as representations of my view?], but what I think is particularly admirable is his lack of bias and the very serious effort he made to see the new developments in relation to the general epistemological position which he himself had formed as a result of extensive studies of psychological problems over a period of many years. As for my own opinion, I would like to add that what the theory of relativity and quantum mechanics have taught physicists about unpredicted possibilities and what it has done to broaden our attitude to the problems of existence and to emancipate us from narrowing frameworks, is something which has relevance for intelligent people in every sphere. In a few days time, when I am back in Copenhagen, I should like to send you some offprints on such topics.

With kind regards,
Yours sincerely Niels Bohr

In this letter Bohr tells Sorainen that he had had many deep and searching discussions with Høffding on recent developments in quantum mechanics, and that Høffding, in these discussions, did not regard the new situation in quantum mechanics and the idea of complementarity as alien to his own epistemological ideas, which had emerged as a result of his work in psychology. By saying that he admires Høffding's whole-hearted efforts, Bohr grants that there was a connection between Høffding's views on psychology and epistemology and also that his own notion of complementarity did not clash with Høffding's theory of knowledge. Indeed, Bohr also adds that Høffding's statements should not be taken too literally as, I would guess, a representation of Bohr's idea of complementarity in quantum mechanics, owing to the fact that Høffding did not understand much of the mathematics involved. But this factor does not prevent Bohr being influenced by Høffding. The point is that by saying that "isolated statements in Høffding's essay cannot be taken too literally" Bohr did not want Sorainen to read too much of *his* influence into Høffding's text. Thus, I shall treat the recognition of these facts in a little more detail in order to show that the notion of complementarity can be seen as a consequence of the application of Høffding's epistemology to the problem of measurement in quantum mechanics.

3. HØFFDING ON COMPLEMENTARITY

In his last essay on epistemology, to which I have alluded in the above section, "Bemærkninger om Erkendelsesteoriens nuværende Stilling" (Notes on the Present State of the Theory of Knowledge), Høffding considers the development of quantum mechanics to be in complete alignment with his philosophical and psychological ideas. When he refers to the notion of complementarity, he does so because he realizes that this idea has its analogue in his own exposition of the characterization of the problem of description. The problem of description arises in connection with the attempt to establish a distinction between subject and object, a distinction which can be drawn only if the object can be subsumed under a causal sequence. It is obvious that Høffding perceives Bohr's interpretation of quantum mechanics as a confirmation of his own doctrines – a point which may be illustrated by the following statement of his:

> When we called this connection or interdependence [between the rational and the empirical element in natural science] complementary, *we are not using a concept which is new to philosophy*; the author refers to the relation between "the consciousness of the freedom of the will and the requirements of causality" and thereby to the relation between ethical and psychological characterizations of one and the same action, ... [a relation] which recurs at every stage at which psychical concentration and the search for understanding gradually prevail.[31]

We shall return to this point below. Let us first look at how Høffding interpreted the theory of relativity in a continuation of his own epistemology.

It was Høffding's view that science had upheld the "static" or "naive" concept of truth right up to the end of the last century. This is manifested in Newton's teaching on space and time, in the teaching of physicists and chemists on atoms, and in the concepts of species in natural history. In these treatments a concept is taken as expressing the absolute nature of existence, although all concepts came into being as a result of thought and for the use of thought.[32] However, it was a central feature of Høffding's philosophy that every concept expresses a relation.[33] Einstein made a similar point, according to Høffding, when he asserted the fundamental importance of the concept of relation for the analysis of concepts such as space, time and velocity. In science it has proved necessary, remarked Høffding, not only to be faithful to the content of observations, but also to take into account the conditions under which the observations were made.[34] Høffding's opinion seems to be the following: in as much as Einstein in his analysis of time, space and velocity found that these concepts express relations, this implies at the same time that the *knowing subject* is no longer left out of consideration in the description of nature. There is nothing given in our experience that corresponds to the ideas of absolute space, time or velocity. Any statement in which the words "space", "time" and "velocity" occur significantly has meaning only when we know in relation to what coordinate system it applies and, in particular, whether this system, even relatively speaking, is at rest or in motion.[35] For Høffding, this is analogous to taking into account the comprehending subject and this subject's perspective, because two observers moving in relation to each other may conceive two events as being, respectively, simultaneous and separated in time if the events are not causally connectable. Each individual observer thus conceives time in a way that is subject to constraints imposed by his circumstances, and hence nobody can claim that his particular conception is more correct than that of others.[36] So, as early as 1921, Høffding saw the theory of relativity as exemplifying one of the most vivid and central ideas of his philosophy: no objective knowledge without a knowing subject, that is, no knowledge claims about objects are unambiguous unless the relationship to the "subject" is specified.

Of course, most physicists will say that this interpretation is a confusion. In a classical Cartesian spectator theory of knowledge the "knowing subject" is a non-physical mind. But Einstein did not address this issue at all. When the spatio-temporal locus is made "relational", the relata are physical things. What changed was the concept of which properties of the object are "possessed" and which are relational. Thus the theory of relativity provides us with a metaphysical, not an epistemological, lesson. Nevertheless, Høffding saw an epistemological lesson underlying the metaphysical one, regarding the epistemology theory of relativity as being in conflict with the epistemology behind Newton's notion of absolute space and time, according to which these exist in a way empirically inaccessible to the knowing subject. For Høffding, the theory of relativity vindicates the subjective character of all physical phenomena and in fact, as we will see, Bohr followed Høffding in adopting this view.

Later on Høffding acknowledged that a similar problem of objective description was to be found in quantum mechanics. In "Bemærkninger om Erkendelsesteoriens nuværende Stilling" he points out that the dependency relation between the knowing subject and the object known that he had analyzed years before is thus of the utmost importance for understanding the crux of the developments in atomic physics at the beginning of the present century. This is no coincidence since we find that similar views were held by Niels Bohr, who, I am claiming, arrived at his view by applying Høffding's analysis of the subject-object problem to the interpretation of quantum mechanics.

It was Høffding's view that the relation between subject and object came to the fore in a new and decisive way in quantum mechanics. Thus he wrote in connection with the "complementary" description of elementary particles:

What is involved is the notion of alternating viewpoints which elucidate the same event not just different aspects of the event. With a certain connection hereto, we find that if we are to understand the processes inside the atom, the relations between the knowing subject and its object can even less than otherwise be disregarded. In any act of cognition a direct or indirect influence is exerted by the subject through which the object that results from the action takes on characteristics different from those that it would have had without the influence. Intervention of this nature may to a certain extent be disregarded in the case of the usual objects of observation, but in the case of the processes of the parts of the atom it is of crucial significance.[37]

In a later context, Høffding wrote:

The demonstration now given of the limits of our usual forms of thought and perception will be of importance for the entire treatment of the problem of epistemology.[38]

The limitation of the forms of thought here discussed by Høffding is a reference to the well-known circumstance that in quantum mechanics it is impossible to describe a causal sequence in space and time for an elementary particle. In this situation we seem to be faced by an ineluctable irrational element of which we, in our continued attempts to cognize reality in quantum mechanics, have to take account: in the atomic world even the "criterion of reality" seems no longer applicable. But this is only partly correct, according to Høffding, in as much as what it impresses upon us is precisely the need to limit the use of the basic concepts that have been applied for hundreds of years. This in turn depends on the difference that exists between the principle of causality, the intellectual expression of the desire for continuous relations, and the concept of causality that prevails at any given time. Høffding points here to a distinction between the principle of causation and the concept of causation, a distinction which he had drawn many years earlier. In quantum mechanics the principle of causation is thus a case of applying a statistical description instead of the coherent causal description of classical mechanics – and this is the result of abolishing the exact analogy between the rational and the empirical element.[39] By the expression "the exact analogy between the rational and the empirical element" Høffding alludes to what he elsewhere calls the analogy between the relation of ground-consequence and the relation of causal determination which is at the core of the classical description. But in quantum mechanics a description which determines the probability of a future state as a function of the present state is no longer deterministic, and thereby indicates that the exact analogy between the relation of ground-consequence and the relations of causation is invalidated.

This view thus expressed in quantum mechanics also finds application in philosophy. Høffding pointed out in his essay that he had earlier called attention to the existence of corresponding *complementary* descriptions in the relation between psychological and ethical conceptions of the same action. Recall Høffding's statement, quoted at the beginning of this chapter, in which he refers to the complementary relation between the experience of free will in the performance of an action and the demand for a causal explanation in the psychological description of the same action. Here again we face the need to make a clear distinction between the rational and the empirical element because, although human actions can be made to conform to psychological laws, no explanation is thereby given of the freedom of choice of the individual. Hence this gives rise to the discrepancy between the causal mode of description of the actions of a person and his impression that he has been free to choose the actions as his own.[40] As we shall see later on, this incompatibility is a manifestation of "the antimony between involuntary experience and reflection" which appears everywhere in psychology, where the individual cannot be regarded merely as an object.

So absorption in the performance of an act and reflection upon its nature are, according to Høffding, two opposite working processes which, in spite of their incompatibility, also complement each other. In a footnote to the 1930 paper Høffding expands upon his reaction to the idea of complementarity by saying that a relationship similar to the one which exists everywhere where such absorption and reflection are in operation may also occur between psychological and physical phenomena: "Here too a complementary relationship may be a possibility – a question I have returned to in several books".[41] We shall return to Høffding's complementary solution to the psycho-physical problem later on.

It is evident from "Bemærkninger om Erkendelsesteoriens nuværende Stilling" that Høffding considered Bohr's interpretation of quantum mechanics as something growing out of his philosophical and psychological ideas, and thus as a confirmation of his earlier analysis of the relation between subject and object:

> Not the least significant aspect of the new investigations in physics is that they remind us of the great problem of the relation between subject and object in epistemology.[42]

He also stresses that it serves to confirm what has been achieved in this area that in quantum mechanics only those views are applied that are found in a similar form in philosophy.[43] Recall also his letter to Meyerson, dated as far back as 30th March 1928, in which he refers to Bohr's Como paper as published in *Nature* ("The Quantum Postulate and the Recent Development of Atomic Theory") expressing a similar view about the roots of complementarity:[44] "Here he tries to overcome the difficulty which lies in the fact that the electron has simultaneously to be a particle which is located at a definite position and a source of energy. Here we have an old problem presenting itself once more at the frontiers of the natural sciences". Bohr is in direct agreement on this point, saying that because of the finite magnitude of the quantum of action, it is impossible to distinguish between the behavior of the atomic object and the means by which it is observed, and we are therefore faced with the old philosophical problem concerning the relation between subject and object which, according to Bohr, is the core problem of epistemology.[45] Let us therefore look more closely at Høffding's philosophy and psychology to see what his position was on epistemology and ontology.

Chapter IV

1. HØFFDING'S THEORY OF KNOWLEDGE

Høffding was largely an eclectic philosopher, although he was not a man who only put new wine in old bottles. Høffding himself pointed to Comte-Spencerian positivism as his philosophical starting point,[1] as we have noted, though his philosophy showed the clear influence of Spinoza and Kant in places. According to him, philosophy has been, ever since Kant, an investigation of the conditions of human thought,[2] and "the epistemological problem arises when it is asked on what the validity of our understanding depends, and how far it extends".[3] But as the ability to understand is integral to human thought, cognition depends not only on the nature of the phenomena, but also on our entire intellectual structure and organization.[4] "All principles and hypotheses will be of a certain type which ultimately point to the innermost nature of conscious mental life. And here one will always return to the need for unity and continuity".[5] By unity and continuity Høffding means something like the connectedness of phenomena in forming a unified conception of nature as a whole. Nonetheless, even if every principle and every maxim can be traced back to this need for continuity and unity, Høffding well realized that these are not thereby proved to be objectively valid. In his opinion all that we have proved is that the principles in question are psychologically possible, because they agree with the general laws of conscious mental life – which is a necessary condition of any understanding at all but not a sufficient one.[6]

Høffding believed that the ultimate task of cognition is to synthesize and compare, which means that the most fundamental categories cognition makes use of are synthesis and relation. These two categories pave the way for a series of other basic categories which appear in pre-scientific as well as scientific thinking. The formal categories (identity, ground/consequence and others) underlie logic and mathematics, and the real categories (cause; totality; evolution) underlie all the empirical sciences, and lastly the ideal categories (the concepts of values) are manifested in aesthetics, ethics and the philosophy of religion. The transition from real to ideal categories is, according to Høffding, characterized by every evaluation implying a relation to a totality, whether it be constituted by an individual, a group or a society.

Paralleling Kant's transcendental deduction, Høffding argues that the various categories of cognition emerge from an analysis of different kinds of judgments. By examining the predicates that are among the most common in our judgments we may characterize the forms of thought in which the items or the phenomena appear to us. But opposite to what Kant thinks this does not make them *a priori*, since the logic of concepts is not fixed, according to Høffding. He holds the pragmatic view that the categories of cognition reflect the need to synthesize experience, and are thus relative both to our needs and to the experience which must be synthesized. However, it is not through judgment that the items of the experience first appear to us. Any judgment presupposes perception or intuition, that is, sensation, memory and imagination. According to Høffding, the items (which is Høffding's own neutral word for phenomena) are ordered spontaneously in and for our mind in intuition or perception ("*anskuelsen*").[7] A form of wholeness is involuntarily attained in the intuition, before consciousness or reflection is capable of establishing any further order among the items and hence making judgments. In other words items are things or events, physical or mental, which are experienced as immediately given wholes. They from the content of perception which exists as impressions in consciousness.

The forms of perception and the categories of cognition together form the requirement of continuity. As indicated earlier one of the main themes in Høffding's philosophy is a dualism of continuity and discontinuity which he claimed underlies every philosophical problem. In his *A History of Modern Philosophy*, when writing about Kant, he gave a characterization of the law of continuity:

> The law of continuity (which includes within both the laws of continuity of space and degree and the law of the causal relations of all phenomena) is valid for all phenomena, because it formulates the general conditions under which we can have real experience (as distinguished from imagination) ... Only as the condition of experience has the law of continuity (including the causal law) validity. ...[8]

A little later in the treatment of Kant Høffding criticizes him for making a sharp distinction between the forms of intuition, the categories of understanding and the ideas of reason, saying that

> ... continuity, causality, time and space – as conceived by Kant – possess an ideal perfection to which there is no corresponding experience. Continuity is an idea to which experience only gives us approximations. What Kant calls forms are, as a matter of fact, abstractions and ideals which, in accordance with the nature of our knowledge, we set up and use as measures and rules for our inquiries.[9]

Thus Høffding regarded the forms of perception as "abstractions and ideals". They have "ideal perfection" to which nothing in our sensory experience corresponds.

Now, the problem of epistemology bears essentially upon the relation between these categories of cognition or forms of thought on the one hand and the items which are the content of experience on the other. In Høffding's opinion, the factor that establishes the validity of cognition is the greatest possible degree of connectedness among our impressions and our ideas, i.e. the ability to represent a coherent, causally connected, unified world. Høffding terms impressions and ideas the elements of cognition, where ideas in their simplest form are to be understood as reproduced impressions. Observation includes both impressions and latent ideas in contrast to recollection, which consists exclusively of free ideas.[10] So, according to Høffding, truth does not consist in a correspondence between a certain idea and a state of affairs in the external world. For as he says,

> Truth cannot be defined as the agreement of our thoughts with reality. We only have knowledge of reality through continual efforts to make the items conform to our forms of thought. Reality, the truth of the items, already consists in practice, for the sound human intellect, in a close connection between as many accurately comprehended items as possible.[11]

Truth is determined by Høffding, purely formally, as the greatest possible unity between our ideas, acquired by consistent reflection from a certain viewpoint.

In his lecture at the Jowett Society in Oxford on 26th November 1904 he put his thought about the concept of truth in the following way:

> The right to establish something as a principle is founded on the claim that it leads to the recognition of a connection between our observations which would otherwise be obscure and sporadic. The truth of principles, then, does not consist in their conformity to an absolute order of things: – an order of things we do not know of before finding – with the help of these principles – a connection between our observations. We ourselves produce the truth, when we find the principles, which can connect items to the greatest extent and to the highest degree. A critical or dynamical concept of truth is in the making, opposed to the dogmatic concept of truth which can be designated as static, since it presupposes a given quiescent order of things which is then to be reproduced in thought. This is nothing new for the philosopher. Critical philosophy had already postulated a dynamical concept of truth, when it pointed out that objective validity consists in the lawful connection of our observations.[12]

By the term "critical philosophy" Høffding alludes, of course, to the philosophy of Kant. So following Kant, this formally determined concept of truth, which Høffding calls the dynamical concept of truth, is to be the basis of the concept of reality, and Høffding thus defined reality as the maximum of lawful connections and the highest possible degree of concord between as many different viewpoints as possible. The criterion of reality thus consists, then, in the greatest possible agreement and lawful connections between our impressions and our ideas,[13] or as he also formulated it, of "the constant connection between our observations".[14]

Although Høffding has here presented us with a formal definition of truth, he characterizes this concept of truth as a "dynamical" one because it emerges from the activity of the intellect, an idea which is kindred to a pragmatic notion of truth. The static concept of truth, according to which truth consists in a correspondence between reality, that is, some state of affairs independent of our mind, and our cognition, has to be waived, partly because it is inconsistent, since we only know reality through our cognition, and partly because we in the process of reflection would never be able to reach a point at which our cognition with its forms and the items themselves could be compared. The only possible concept of truth is the dynamical one. In the lecture he gave at Harvard, when he visited William James in 1904, he stated:

At the present time there is a growing consensus that the significance of scientific and philosophical principles consists in the guidance they give us in our striving towards understanding. Their truth is their validity, and their validity is experienced through their capacity to guide us in our intellectual endeavour. A principle is true if it can be used, that is, if we can work with it to gain knowledge with its aid.[15]

In another context Høffding expresses a similar view: "The truth of principles consists in their validity and their validity in their *usefulness*, their ability to propel research forward".[16] Consequently, Høffding regards the usefulness of scientific principles and theories as the ultimate criterion of their validity. And their usefulness is revealed by their capacity to bring together as many phenomena as possible in an invariant way.

Thus, Høffding's view of truth seems to be something of a hybrid of truth-as-coherence and truth-as-usefulness. For him a statement is true if and only if it can be connected in a consistent way with other true statements. Moreover, these other statements must be such that they form the most comprehensive system of statements possible. This is the coherence component of his notion of truth, which deals with isolated statements.

But Høffding seems to have acknowledged an obvious difficulty facing the coherence theory, even if he did not mention it, viz. that the truth of the system of statements cannot itself be grounded in the same principle. However, Høffding seems to have found a solution to this problem in that he combines the idea of the truth of individual statements as coherence with a pragmatic theory of truth for a system of statements according to which the system itself is true if and only it can prove its usefulness by guiding our research and explaining new phenomena alongside of well-established ones. So we produce truth, according to Høffding, in the sense that it is we who formulate the principles of knowledge which prove capable of coherently uniting as many individual statements as possible, just as it is we who determine whether the principles are fruitful from a scientific point of view.

The criterion of reality consists, as already noted, in the lawlike connection between our observations; but such a connection can only be demonstrated if it can be established that there exists a causal connection; that is, if it can be shown that two phenomena are so linked together that when one is given, the other inevitably occurs.[17] This means that there lies a problem in Høffding's use of the concept of "lawful connection". The problem is that the concept sometimes refers to the nomological laws of a theory and sometimes to the laws of nature themselves. Thus the concept partly covers what Høffding called the relation between "viewpoints", i.e. logical ground-consequence relations, and partly the relation between the items with which the "viewpoints" are concerned, i.e. the causal relations. Høffding failed to make any distinction, apparently because he regarded these two relations as analogous, the analogy itself serving to determine the function of causality as a criterion of reality. However, considering that the ground-consequence relation can never, according to Høffding, be identical with the causal relation, and considering that he perceived the ground-consequence relation as identical to the "lawful connections between the viewpoints", it is improper to speak of the "lawful connection among our impressions and ideas" at the same time; because causality is, namely, the criterion for the existence of a "lawful connection between the viewpoints of consciousness". Thus our ability to establish causal connections between the phenomena determines the validity of the nomological representations of phenomena.

Nevertheless, Høffding thinks that causality is reflected in *consciousness* as a real unity through its there creating uniformity and continuity between apparently diverse phenomena.[18] But this also implies that something can be recognized as real – as being independent of our subjective impressions and ideas – only in so far as it enters into a causal connection. Recall Kant's well-known example. The sequential order of our experiences when we are looking at a house is something which we may determine ourselves, in contrast to the sequential order in which our experience of a ship sailing down the river is determined. In the latter case the order of our experience is forced upon us by the existence of a causal connection between various states of the boat. This example illustrates quite well what Høffding had in mind. He maintained, in contrast to Hume, that things are always given us as part of a connected sequence. In general, he holds that "we only have knowledge of a thing in so far as it is a cause or an effect".[19] This means that the only knowledge that we can have of a thing is of its connection and interaction with other things. Moreover, all that we know of a thing is its properties, and the properties of a thing are nothing more than the ways in which it affects or is affected by other things.[20] Thus properties are relational, not possessed. The color of an object, for example, is the manner in which it reflects light, and its hardness is the resistance it makes to a penetrating body. This definition of property exemplification also applies to particles, the smallest parts of matter. "What we call 'things', immediately given wholes, are only understood qua their properties, and the properties make manifest just as many relations to other 'things'. Molecules, atoms and electrons are still 'things', totalities, only known and understood qua their relations", Høffding says.[21] The result is that we are unable to ascribe any properties to things that have no causal relations to other things – hence we cannot have knowledge of them and they will seem not to exist for us.[22] Consequently, Høffding considered every property to be relational.

The *principle of causality* cannot, Høffding emphasized, be proved to have universal applicability. It can only be regarded as a hypothesis, in as much as we cannot prove that all phenomena can in fact be traced back to other phenomena as their causes. The *concept of causality* is nevertheless intimately connected with the intrinsic functioning of the mind and expresses its search for connectedness and unity by relating impressions and ideas. Thus, by following the law of causality as a principle, we are aided in establishing a continuous connection between phenomena and thereby an understanding of them.[23] Neither can it be proved, therefore, that the criterion of reality is applicable to all items or phenomena; we have in it merely a "means for thought and a form of thought that we must continually attempt to apply, if only for the reason that we would otherwise be without a source of orientation in our dealings with the world".[24]

It has been claimed that Høffding, apart from considering the principle of causality a defining feature of reality, considered it a research principle.[25] Neither part of this statement seems wholly correct; and even though Høffding is not always entirely unambiguous in his writings, he does seem to have been clearer on this point than one perhaps might have expected.

Firstly, Høffding was not merely talking of the principle of causality; on the contrary, as we have just seen, he distinguished between the *concept of causality* and the *principle of causality*.[26] Secondly, the concept of causality is directly connected with our ability to perceive a reality independent of ourselves, in virtue of all our impressions and ideas. Reality is defined by Høffding as "the lawful connection and agreement between as many different viewpoints as possible". That is, a given number of "viewpoints" or assumptions about some items are true and belong, if at all, to the realm of the real only if they can be brought into connection with the other assumptions of the knowing subject. So the concept of causality imparts to the knowing subject a form of thought which, combined with other concepts dealing with the items existing for consciousness, gives rise to a stable and continuous connection between observations, and this connection constitutes the criterion of reality. Therefore, if the items can be subsumed under the concept of causality, then consciousness will have demonstrated a causal relation and thus a lawful connection between them, and we have the right to conclude that the items exist independently of us. The concept of causality was thus, for Høffding, an *epistemological criterion* of that which is real, and it cannot in itself constitute any *defining* feature of reality. Høffding would probably raise the objection that the conception of causality as a defining feature of reality would make the concept of reality into a purely *ontological* concept – for which he criticized Kant. For, as Høffding maintained, "Only for thought does reality exist".[27] The principle of causality, the principle "that everything has a cause", concerns, on the other hand, the general and universal applicability of the concept of causality. Just because the use in thought of the concept of causality is the condition for every empirical cognition does not mean that we have demonstrated that every phenomenon existing for consciousness can be subsumed under this concept. And we cannot show this either, asserted Høffding, as he claimed Hume had already quite correctly pointed out. Therefore the principle of causality, which is in itself a hypothesis, expresses a principle that gives guidance about any task facing the knowing subject, i.e., it leads our efforts to find a causal relation where we cannot immediately ascertain one. Høffding maintained that the principle of causality is the intellectual expression of the need for continuity. Later we will be introduced to domains of experience that Høffding considered could not be made to conform to the concept of causality, such domains thus representing an irrational obstacle for cognition. Høffding's views on the present point can be summarized as follows: The principle of causality cannot be shown to have universal applicability, but cognition of reality is only possible provided that the content of consciousness is amenable to subsumption under the concept of causality; in other words, causality is the criterion for what is real.

But the claim that a thing is real is not a certainty once and for all. Høffding asserted that the concept of reality is still in embryo, as it is continually being altered through the use of the appropriate criterion in force at any given time.[28] The notion of "existence" or reality is in other words an ideal concept.[29] This is because the nature of the connections found in reality are determined by the basic concepts and categories that prevail at any given stage of development of science, and which form the conditions upon which all research rests. Hence, the acquisition of human knowledge will always be tentative and will never end. Thus Høffding wrote, "Even pre-scientific consciousness used certain basic concepts spontaneously (I call them the fundamental categories), which science, as it developed, elaborated according to the requirements of the items".[30] The history of science has at the same time shown that these categories of cognition are not unalterable, but that they can come into being, alter or become obsolete.[31] This view stands in stark contrast to that of Kant, who believed that the categories of the understanding were given once and for all. Høffding considered substance to be a perfect example of a moribund, if not already defunct category. The validity of categories and principles lies therefore exclusively in their operational value, and they are only valid as long as, and to the extent that, they can be used to yield understanding.[32]

However, Høffding thought it possible to discover three pairs of fundamental categories: synthesis and relation, continuity and discontinuity and similarity and difference, which, he asserted, manifest themselves in all conscious mental life, and which are thus common to pre-scientific, non-scientific and scientific knowledge alike. The so-called formal and factual categories in, respectively, logic and the empirical sciences are particular and more accurate determinations of these fundamental categories.[33] In Høffding's own words, "The particular development of the forms is determined by the tasks required of thought. Therefore a distinction is made between fundamental categories on the one hand, and the formal, factual and ideal categories on the other".[34] Thus it is only the last species of category that we can be forced to alter, limit or entirely relinquish, all depending on whether the principles in which they come to expression appear unable to make experience conform to them.

Naturally, the question arises of what the relationship between these categories of cognition or forms of thought, on the one hand, and the items, the phenomena, that are the content of cognition, on the other, consists in. Høffding wrote that "only by means of thought can we justify our belief that we are confronted with a reality". But, he argued, this fact "cannot be interpreted so as to lead to the result that this reality exists only in the forms of thought themselves. The relation between observation and interpretation can never be identity".[35]

In so far as factual categories are concerned, they are the product of experience.[36] The categories express the way in which thought functions given perceptual interaction with the items. Through them we gain knowledge of our intellectual organization. What we understand – and whether we understand anything at all – depends both on the nature of the items and on the nature of our minds; just as which colors we see and whether we see any colors at all depends not only on external objects but also on the nature of our organs of sight.[37]

In other words, Høffding maintained that the items are not given *per se* to our consciousness but are themselves the result of a process. The modification that the items are subjected to is a necessary condition for the existence of any form of cognition at all. But even if our observation depends on the categories of cognition, this does not mean that the items cannot influence these, in as much as the more specialized categories prove inadequate for the purpose they are to serve.[38] According to Høffding, there is a constant interaction between the items and the forms of cognition which results in the diversification of both items and thoughts. Hence observation and interpretation are not one and the same. He emphasized this point in the following elegant statement of the empirical underdetermination of theory by the phenomena.

> The dynamic concept of truth does not obliterate the difference between item and form. On the contrary, it impresses upon us that the way in which the items have hitherto been treated in science cannot be proved to be the only possible treatment. It has been possible to put forward principles and hypotheses on the basis of which the items can be linked in virtue of stable and exact connections, but we cannot prove that the principles and hypotheses that we put forward are the only ones possible, the only means of reaching the same result. The success that science has enjoyed is no proof of its absolute truth so long as it cannot be proved that no other assumptions could have led to the same result.[39]

However, since we can only match thoughts with phenomena, and not cognition itself with an absolute order of things, according to Høffding, the cognition of an absolute order of things is in principle impossible, contrary to the tenets of the proponents of the static concept of truth. Høffding reproached Kant for what remained of his dogmatic slumber in his use of a "Ding an sich", existing absolutely independent of any cognition. Yet it was not only Kant who still in a certain respect embraced the static concept of truth. Even the natural sciences had done so up to very recent times, according to Høffding. In 1910 he wrote:

> This is particularly clear in the case of Newton's doctrine of space and time, in the physicists' and the chemists' doctrine of the atom and in the naturalists' concepts of species. Just as Pythagoras posited numbers, Plato ideas, Spinoza substance as expressive of the

> absolute, so too in this way space, time and atoms were posited as an expression of the absolute nature of reality, despite the fact that these very concepts are the product of, and for the use of, thought. From their point of view it was to be deemed a failure that we were unable to attain to the pure world of numbers, ideas or atoms or to absolute substance or absolute time and absolute space. Critical philosophy, on the contrary, regards all such concepts as means and forms of which we avail ourselves while working to attain the perfect connections we call reality.[40]

Høffding then added that "reality" is not an otiose byproduct of the pure world of ideas, but through the work of thought, the ideas guide us forward towards more consummate concepts of reality.

Høffding asserted that in the very relation between forms of cognition and phenomena – with which we have just dealt – lie, at one and the same time, the conditions for our cognition and the limits to our faculty of cognition. This is so because any single concept acquires content through its relations of similarity and dissimilarity with other concepts. All the concepts with which we operate in cognition prove to be relative, i.e. they express relations, and they can therefore be applied only to that which can be considered part of a relation. We cannot, as stated earlier, have any cognition of that which, given its nature, cannot have a relation to something other than itself.[41] "Cognitive activities still concern the finding of new relations and the search to make them correspond with relations found earlier, or perhaps to use them to correct these".[42] Thus it serves both as a condition of, and a limit to, our cognition that everything of which it is possible to have knowledge must be related to something else. Moreover, this means at the same time that the process of cognition never comes to an end, it being continually possible to uncover new relations and correspondence between items where this has not been done already.[43]

Nevertheless, this cognitive limitation, as formulated by Høffding, results in a more significant difficulty. Høffding propounded the view that an *irrational* relation will permanently persist between the concepts or forms which our consciousness is capable of creating and reality itself, from which our experiences originate. Because of the *contradictions* arising from our experience, a full realization of the ideal of cognition, which would be that of the universal unity of and connectedness of all items, must be assumed to be impossible.[44] Høffding's line of thought seems to be that in as much as the criterion of reality is restricted to that which can be brought under a constant and regular connection between items, and furthermore that our consciousness does not succeed in bringing about such connectedness in all areas, the human intellect will continue to be confronted with an irrational element which does not conform to the categories of the understanding. The rational element for cognition is, of course, the continuous connection between the items. According to Høffding, this means that anything that cannot be made to conform to a causal connection, such as is required by the faculty of cognition, will remain as an irrational element for cognition. Hence not every element of experience can be made to conform to our comprehension of reality.

In an essay entitled *Charles Darwin og Filosofien* (Charles Darwin and

Philosophy) Høffding points out that the inability of the understanding to subsume under a category every item or phenomenon in the mind is a fact which has been overlooked very often in epistemology. Cognition does not always have forms of thought at its disposal with the aid of which new items or phenomena become intelligible for the knowing subject. As he puts it:

> The question is whether thought has forms under which novel experience may be subsumed. We are here touching on one of the basic conditions of knowledge which even the great masters of epistemology did not elucidate. Kant took it for granted, with seeming assurance, that only those entities which could be rationally ordered in our forms of cognition are such as are able to issue forth from the obscure spring he called "the thing in itself". He tells us of the possibility of other forms than those pertaining to the intellect of man, and he warns us against the dogmatic assumption that human apprehension of reality is absolute and exhaustive. But he seems to be fairly confident that "the thing in itself" operates constantly and consistently and invariably gives us that which lies within our ability to handle. This condition can be accredited to Kant's rationalistic tendency; but no arguments can be given to support it.[45]

In opposition to Kant, Høffding believed that "the given" (the items) under certain circumstances imposes restrictions on the rationality of cognition in the sense that in its attempt to systematize continuous connections, the cognitive faculty is sometimes confronted with irrationalities or incompatibilities in the form of real discontinuities between the forms of thought and the items, or as he also says, between the rational and the empirical elements of our cognition.

Høffding claimed to be able to show that there were (at least) three such essential incompatibilities or irrationalities. In the first case, qualitative differences give rise to a lack of convergence in so far as they cannot be reduced to quantitative differences. Qualitative differences do not become less real by being termed "subjective qualities" – this will only postpone the call for explanation. Qualities are straightforward facts and there is no explanation of how it might be possible for purely quantitative differences to appear to our senses as qualitative differences. Attempts to substantiate the claim that such an explanation obtains are tantamount to making the analogy between quantity and quality into an identity, asserted Høffding.[46] Since, then, there is no identity between them, the causal relation will not be "clear and transparent when there continue to be differences of quality between the event standing as cause and that standing as effect".[47] Differences in quality thus constitute a limit to complete proof of a lawful connection.

In the second case, there is an incompatibility in so far as temporal differences cannot be eliminated or reduced to a formal identity relation. At most, a series of events can be so ordered that there is an analogous relation between the real and the formal categories, between cause-effect and ground-consequence, which is necessary to achieve an understanding of reality.[48] As Høffding also puts it, "In the logical relation between reason and result, differences of time and quality play no part, and if the analogy is to hold such differences between the items must be reduced to the least possible".[49] But there can never be any question of an identity because the cause-effect relation distinguishes itself from the ground-consequence relation by the fact that the former involves a succession in time. If, nevertheless, one claims that there is such an identity, then one simultaneously does away with the concept of causality, which is the very concept that makes possible the empirical sciences. But, on the other hand, this means that we cannot eliminate the passage of time, even though "thought is still concerned with a rationalization of the items through setting up series that deploy the concept of identity to an increasing extent".[50] In the causal relation the time sequence is the irrational element that cannot be reduced to anything subjective or be entirely eliminated, and which prevents complete cognition, because at all times we can only have knowledge of the past and can merely conjecture about the future.[51] The concept of causality is, however, at the same time the condition for our possession of any knowledge at all, even though we cannot prove that the law of causality is universally valid. Hume's problem of induction still remains unsolvable, according to Høffding.

Finally, in the third case, there is the disparity between the knowing subject and the known object. As Høffding put it in his *Filosofiske Problemer*, "In all acts of cognition, it is possible to distinguish between a subjective and an objective element – but both elements are given only in relation to each other, although in the context of this relation they can manifest themselves to different degrees".[52] This statement raises the question of what is subjective and what is objective in our cognition. This cannot be answered, asserted Høffding, by making reference to the fact that the world is external to or independent of consciousness, any more than by referring to the fact that all qualities are given *eo ipso* our consciousness.[53] For it is the very qualitative differences that, for our cognition, constitute the given material and provide the matter of enquiry. Consciousness itself cannot produce the differences that are the material for its activities, whereas the form in which, and the degree to which, they appear to consciousness are determined by the invariably functioning conditions of consciousness itself. Our comprehension thus strives to transform the actual, existing differences into stages of one and the same continual process, or to forms of one and the same content.[54]

This leads, thought Høffding, to the circumstance that

> ... when we distinguish between subject and object in our cognition, we contrast in fact an objectively determined subject with a subjectively determined object. The properties that we ascribe to the subject cannot be explained on the basis of the concept of subject (the pure subject) itself; they stand as facts just as well as all the other properties with which our cognitive faculties deal. And the properties or determinations that we ascribe to the object are only attributable to it in relation to a subject and, defined in greater detail, to a subject of a certain, particular nature.[55]

So whenever we regard something as an object we have to determine the nature of the subject in relation to which it manifests itself, and whenever we regard something as a subject we have to investigate the objective connections which determine its nature or its relations to the object and which make the subject an object for another subject. Thus, Høffding believed, there is never a "pure" object, but only an object that is comprehended by a subject and colored by it, and likewise there is never a "pure" subject, but only a subject whose nature is partly determined by the objects that constitute the world that surrounds it.[56] He believed, in other words, that a subject has to be specified in relation to its objective content, the items, since the content imposes certain constraints on the subject's application of the forms of thought, just as the characterization of the objective content is dependent on the structure of subject, viz. its *current* forms of perception and thought. However, there seems to be a difference between the epistemic status of the object and of the subject because we do not know the surrounding world *per se* but only as it appears to us. As Høffding said, "The subject is the Archimedean point in the theory of knowledge – the point from whence reality can be, not moved, but experienced".[57] More than what appears from such a point cannot be known, and neither can it be known how it might otherwise appear. From an epistemological point of view, the concept of consciousness underlies the concept of matter because all that we know of matter we know through our consciousness, and we only have direct knowledge of the content of our consciousness.[58] Thus, in a different context, Høffding wrote that "just like sense qualities, space, objects and cause, 'reality' is a predicate that the knowing consciousness, from its viewpoint or according to its nature, ascribes to be objects".[59]

The dilemma arises each time we ask from whence the subject gets its objective (actual) properties, and what relation obtains between the subjective determinations of the objects and its nature. Høffding thinks that irrational elements arise because it is impossible to isolate a "pure" object or a "pure" subject. Absolute objectivity and absolute subjectivity denote ideals that we can only perpetually approach but never reach. In reality, subject and object will continue to interact in so far as the cognitive process makes progress, in as much as neither subject nor object can be eliminated nor the one be derived from the other. The irrationality appears because of a continuous series of subjects and objects like $S_1\{O_1\{S_2\{O_2\{S_3...$, in which the preceding element is characterized in terms of the succeeding element, but in which the difference between an instance of the subject and of the object manifests itself once more after it has been possible to give an objective characterization of the subject and a subjective characterization of the object. Irrationalities like this both hinder and promote the acquisition of knowledge, partly making it impossible for such acquisition ever to be brought to an end and partly determining the continued progress made by our search for knowledge.[60]

The continuous development of scientific knowledge is in fact, according to Høffding, ensured by a permanent interaction between what is objective and subjective in the world. Progress in the acquisition of knowledge consists partly in making the objective subjective, that is, in isolating objects for cognition, something which depends on the nature of the knowing subject, and partly in making the subjective objective, that is, in making the nature of the knowing subject an object for science.[61] Høffding points out that we know the world through ourselves but that we also know ourselves through the world. The romantics, he says, particularly Hegel and his successors, developed the former and neglected the latter; and the positivists laid stress on the latter and disregarded the former. However, neither aspect can be eliminated; each is nourished by the other.

In the interaction between subject and object the former is confronted with various pairs of contrasts in its attempt to bring unity and continuity to the latter. The most important are unity/plurality, mind/matter and continuation/evolution. On the one hand, as we shall see, the contrasting components in these incompatibilities cannot be explained away but, on the other hand, neither can they be derived from one another. Hence, they appear as three discontinuities which make it difficult to create a coherent picture of the world.

However, according to Høffding, such discontinuities prevent, inter alia, the progress of the acquisition of knowledge ever coming to an end. The non-finality of knowledge was central to Høffding's thought. One of his students, Frithiof Brandt (1892–1968), professor of philosophy at the University of Copenhagen from 1922 to 1958 and intimately acquainted with Harald Høffding's philosophy, once made the following characterization of it:

> Even those who have not known professor Høffding personally, will often have had occasion to observe the urgency with which he warns us in his writings against all forms of finality. Any limit, any thought of absoluteness, completeness, finality, imperturbability, did not only defy his critical thought, but challenged his sentiments and will as well. He embodies an excelsior, constantly alive, which does not give in or surrender. Somewhere he says: if I had to choose a symbol for my philosophy, I would direct attention to an irrational number, ascertainable with more and still more decimals but, none the less, inexhaustible. As it happened, he cherished or highly appraised Leibniz' symbol, the spiral, which inclines only to resurge. Thus, his constant emphasis on keeping the problems open. Thus, a remark like this: I believe more in ideals than in ideas. Or, in a more abstract and technical manner: I give the tackling of problems preference over their solutions and, likewise, the method preference over the outcome or result.[62]

Brandt then called attention to G.E. Lessing's famous words from "Eine Duplik", which Høffding had quoted with affection in his *A history of modern philosophy*, saying that if God in his right hand had the truth and in his left hand were holding the striving for the truth and offered him a choice, he, Lessing, would choose the left one.

Høffding's principle of the non-finality of cognition was in fact the subject of a speech by Edgar Rubin which was given at the same time as Brandt's to honor Høffding on what would have been his 89th birthday, 11th March 1932. Already in 1882 Høffding had closed his *Psykologi* with the following remark: "The proposition which aims to explain everything becomes its own final – and constant – problem". Thus, Høffding believed that a complete picture of the world in which every item had found its place is impossible. New items are constantly appearing and with them new tasks and challenges.[63] Or, as he said elsewhere, none of the specific sciences is ever complete because new experiences and new puzzles are produced continuously.[64] In this same book he also emphasized that in the theory of knowledge preconceived ideas have to be assessed again and again since one can have no guarantee that one is in possession of the ultimate premisses.

If we look at the historical situation around the turn of the century, in which context Høffding presented his doctrine of the non-finality of cognition, we will see that many scientists at that time thought that all the essential questions in chemistry and physics had been settled conclusively by Newton's mechanics and Maxwell's electrodynamics and that only a few fringes or marginal areas remained for further research. Only a few scientists had the feeling that something was entirely wrong. Hence, it was with an unusually open-minded attitude to scientific progress that Høffding, one year before Einstein started his scientific revolution, was teaching Bohr and other members of the Ekliptika Circle epistemology and scientific methodology.

How important Høffding regarded the principle of the non-finality of cognition can be seen from the fact that to him it was a principle which might be connected with the very nature of all that is. He believed it to be a possibility that, when the acquisition of knowledge knows no end, it is not merely a remarkable thing to us as knowing subjects and our relation to the other parts of the world, but it may be linked up with reality itself – that it is not over but still in the making. Reality has an unfinished and incomplete temporal form. So the source of all irrationality is not merely to be found in one of our categories but also in the form of reality which we have no right to eliminate through explanation.[65]

When Høffding published *Filosofiske Problemer* (Philosophical Problems) in 1902 and *Den menneskelige Tanke* (Human Thought) in 1910 his views on the theory of knowledge and ontology were in all essentials fully developed – long before Niels Bohr's interpretation of quantum mechanics in the late 1920s – and many of Høffding's basic suppositions can even be traced back to the early 1880s, around the time he published *Psykologi i Omrids på Grundlag af Erfaringen* (Outline of Psychology on the Basis of Experience). His later works on epistemological problems, "Totalitet som Kategori" (Totality as a Category), 1917; "Relation som Kategori" (Relation as a Category), 1921; "Begrebet Analogi" (The Concept of Analogy), 1923 and "Erkendelsesteori og Livsopfattelse" (The Theory of Knowledge and Apprehension of Life), 1925 can thus be regarded as amplifications of aspects of his theory of knowledge treated at an earlier time.

2. TOTALITY AS A CATEGORY

Recent research has drawn attention to the fact that Bohr's philosophy contains highly conspicuous features of holism.[66] This is not surprising, since such features were also essential characteristics of Høffding's scheme of philosophy. He rejected a representational theory of knowledge, claiming that the nature of reality is determined by the entire theoretical description of our immediate experience at any given time. But constituting a part of this scheme was a conception of individual items which fail to fall within a causal description of relations between the whole and its parts. That means that his philosophy comprised both epistemic as well as metaphysical elements of holism, or rather, emergentism.

It was Høffding's idea that the most important forms of thought or categories could be ordered in a progressive series of differences: synthesis, identity, rationality, causality, totality and value. Each successive link in this series logically presupposes the preceding links. In his work "*Totalitet som Kategori*" (Totality as a Category) Høffding's intention was to demonstrate that the distinction commonly drawn between the science of nature and the science of what is human based on the assumption that the former aims at finding general laws among its elements while the latter concerns individual wholes is a superficial conception. He believed that the concept of wholeness applies to all items which appear as immediately given wholes, both to thought as well as to the work of human thought. In virtue of its own nature, thought forms wholes in every case where it is possible for the process of reflection to synthesize the items which it had previously analyzed as split up into various elements.

So in science, according to Høffding, we meet two quite different applications of the concept of wholeness. The concept finds a use when the conception of a unity is regarded as the aim of science, in virtue of its search for interconnections between different series of causal connections. After the process of reflection has separated the immediately given, i.e. the items or the phenomena, into their various elements, it synthesizes them once more into new wholes, which until then may have been unapprehended. This happens every time reflection brings about the subsumption of the elements under the concept of causality, either by bringing particular elements into lawful connections or by articulating universal laws which may explain singular causal connections. Thus, new items will continually present themselves to thought, creating new tasks for the cognitive faculty. It is a process which will never come to an end, for what has hitherto been synthesized must forever be synthesized with new elements. In this context we are using the concept of a whole as an idea in the Kantian sense of the word. Høffding holds that this sense is not applicable to what can exist in intuition or perception or to what can be grasped with any finality in thought. Nevertheless, the concept of totality in this sense is a borderline concept. Thus it may not merely be characterized as a specific category or form of thought but as a feature which belongs to each of our categories of cognition or forms of thought. Because synthesis is the essence of our thought, Høffding thinks it is impossible to maintain Kant's sharp distinction between category and idea.

The other use of the concept of a whole applies to what exists in intuition as immediately given items. In that case it is not the work of reflection which produces the idea of wholeness but certain items themselves which invariably appear to us as immediately given wholes prior to any work of reflection. Høffding believes that we do find reference to such items in the field of psychology, biology and sociology. Indeed, there is a strong inclination to find an explanation of the internal and external conditions of these items in the concept of wholeness itself. In psychology people have succumbed to this temptation by letting the soul be a cause of its own states, in biology the same thing has happened when people regard an organism as a being containing in itself the cause of all events occurring within it, and in sociology people have fallen victim to the same error by regarding society as a whole as the cause of particular social phenomena. But Høffding cautions against this tendency.

His position is this: The concept of wholeness can never be an explanatory one. In spite of the fact that a certain item appears as a whole, no understanding follows from such an experience. In virtue of its being a form of thought the concept is necessary for the description of our immediate experience whenever we are confronted with individuals in psychology, biology or sociology. But from that description it is not possible to deduce an explanation of any particular features of these individuals. The whole cannot explain the parts because that would involve seeing the whole as a *causa sui*. But the opposite is not necessarily true either. Although it is one of the tasks of psychology, biology and sociology to attempt to discover the laws which connect the parts and of which the wholeness consists, Høffding argues that there is no guarantee that such efforts will be crowned with success. In "Totalitet som Kategori" Høffding adds to his brief discussion of biology the observation that the impossibility of a physico-chemical explanation of the origin of life cannot be proved, and furthermore that the wholeness of organisms might even be given a mechanical explanation. However, Høffding seems here to have forgotten the lesson of *Den menneskelige Tanke* in his eagerness to set no limits to scientific progress, for there we are told more than once that according to the present sense of wholeness, an irrational relation obtains between the whole and its parts. This relation results in an antinomy arising from the disparity between the rational and the empirical elements in our cognition of the proper domains where experience supplies us with immediately given wholes.[67] This means, of course, that the whole in these cases cannot be explained comprehensively on the basis of a causal connection of the parts. There has to be something left over, even if it is impossible to prove what cannot be accounted for in physico-chemical terms. This follows directly from Høffding's idea of the existence of an irrational relation between qualitative and quantitative differences and from the fact that there are always qualitative differences between a whole and its parts which are not reducible to quantitative differences.

Høffding was of the view that the fundamental will determines in every single case what will be of value, either of ethical value or of any other kind, for an individual considered as a totality. All values presuppose a wholeness of some kind and the value a certain item receives is conferred on it by the whole in relation to which the item is seen to exist or to which it exists. Every value has reference to the relation between a whole and the conditions for its persistence. It is this concept of wholeness which prevents causal descriptions being exhaustive in the sphere of the human sciences as well as in psychology, biology and sociology.

A more detailed account of Høffding's position with respect to psychology will be provided in the following chapter and with respect to biology in a later one. So we shall end this discussion by explaining briefly his opinion with regard to sociology. Since Høffding believed that a society was a whole which could not be conceived merely as the sum of individuals, he also rejected any attempt to see a society as analogous to an organism. He stated his reasons very clearly in his paper on analogy, which he read to the Jowett Society in Oxford in 1904:

> In sociology there has been an attempt to find a leading principle, a fundamental scheme in the analogy between society and organism. Eminent authors have tried to build upon this analogy, but they have not been able to make the image of organism a consistent and fruitful scheme for sociological research. In the first place, society is a connection of organisms, and the relations within one organism cannot be transferred to a whole group of organisms. And again, while in the organism the function of some elements is connected with consciousness, in society all parts are conscious, and must, therefore, be considered as ends in a manner which cannot be said to apply to the parts of the organism.[68]

So, Høffding concluded, the conception of society as an organism is a poor analogy because it does not respect the requirements of fit on all points but only on a single point. However, on the other hand, the idea of society as a whole cannot explain any single social phenomena. To say the opposite would be to surrender to a kind of social mysticism, Høffding believed, in the same way as vitalism and spiritualism are manifestations of a brand of biological and psychological mysticism. The notion of wholeness is a borderline concept which at once refers both to what is experienced as an irrational element for our cognition and to what will be as a permanent challenge for the work of reflection.

3. HØFFDING'S PHILOSOPHY OF MIND AND PSYCHOLOGY OF FREE WILL

Holism, or rather, emergentism, was an essential feature of Høffding's psychology. More than any other empirical science psychology had been the discipline from which Høffding's philosophical reflections sprang. The form which Høffding held that description of psychological experience should take was the channel through which he exercised direct personal influence on Bohr's concept of complementarity and its application in the field of psychology. It is here we are confronted with phenomena, Høffding claims, with regard to which the identification of causal connections cannot serve as the criterion of an objective description because the mind itself is more than the sum of its mental states and dispositions.

The personality or the self manifests itself to us, according to Høffding, as a whole in a dynamic synthesizing process. Like Kant he characterized the mind as a synthesis. Thus, on the one hand, he opposed the assumption that the personality or the self is merely a result of an ensemble of various self-sustaining psychical elements as was claimed by Hume and other British empiricists. On the other hand, he also rejected the hypothesis that attempts to explain every single element of consciousness on the basis of the self or the personality as a whole. His grounds for this rejection were that such a hypothesis uncritically makes the assumption that the self exists quite independently of the elements which it seeks to explain. The correct solution to the problem of what constitutes the individual has to be found in a "law" or a "uniting force". Where Hume merely saw particular self-sustaining perceptions whenever he tried to determine the value of the self, and was therefore unable to find a connecting principle, Høffding identified the self with the forms of thought and laws of connection of sensations, impressions and emotions. The only law which is ubiquitous in the field of psychology is the law of synthesis. This view is familiar from Kant. The law of synthesis determines how all the elements of mental activity in the mind are experienced. Self-consciousness or self-awareness, which is the most characteristic feature of the personality, also presupposes a comprehensive synthesis within which an analysis of the elements that determine reflection about the self may take place. Nevertheless, the elements which are thus identified have to be synthesized before self-awareness can manifest itself.

So, it was Høffding's belief that the self is not definable as an ensemble of all its elements. He thus adopted the non-reductivist standpoint of believing that the self was something over and above its various features. As he formulated it himself, "The synthesis of consciousness cannot come into being merely by virtue of the connection of its individual parts. This means that a spiritual connection is quite distinct from a physical connection, and this is precisely why the life of the mind presents such a great problem".[69] In psychology we are not first confronted with the elements and thereafter with the whole; the given, that is the mental states, appears from the very beginning in the form of a totality, and we are only able to separate the mental states into their various elements subsequent to observation and analysis.

Høffding's notion of the self as the synthesis of mental activity may appear very similar to that of Kant, but there is a fundamental difference. Kant operated with both an empirical and a transcendental ego, while Høffding spoke of a "formal self", something different from Kant's transcendental ego, and a "real self".[70] The formal self is presupposed by any notion of all consciousness. It presents itself to us as the unity of the internal connectedness of consciousness, a unity which is made manifest in memory and synthesis. But we are in principle unable to be entirely aware of the formal self because the mental state which constitutes our thinking of ourselves is always determined by the synthesis. As a presupposition of mental activity the formal self cannot be immediately experienced, but what is known of it is based on an inference from nature and the preconditions of mental activity. In other words, according to Høffding, self-consciousness is only possible by virtue of a process of synthesis, which, when it occurs, gives rise to new experiences which require a new synthesis subsuming the previous synthesis, one that is an experience itself, and so forth. He repudiated Kant's sharp distinction between the categories and the ideas which in Høffding's view cannot be maintained. Høffding was too much of a 19th-century positivist to allow the existence of transcendental entities, which were objects for pure reason only. This means that his notion of the formal self is not absolute but is relative to the state of development of the mind; a new-born baby has a very weak formal self in comparison to that of an adult.

Indeed, Høffding is really here trying to have more than a positivist should have. His idea of a formal self is a sort of pragmatized version of Kant's transcendental ego, but it is still something and it is not entirely an *experienced* something, so it is a sort of transcendental something-or-other. For if it is not something, how can it do anything? Yet its whole purpose is to do something – to synthesize experience.

In addition to its formal unity any individual mind also contains a real unity, according to Høffding. Thus every self is a formal and a real self, but the real self of a person is different from that of another person. The real self is the individual particularity of each single person, that is, the specific content of each individual mind which is synthesized by the formal self which, in contrast to the real self, is common to every person. This content of the mind, the real self, is made up of a stable group of impressions, images, emotions, memories and so on, of which each individual is immediately aware or immediately able to apprehend. Certainly this does not imply that these mental elements have to be present to consciousness at every moment of life but only implies that they have to be capable of being made manifest in the appropriate circumstances. The formal and the real self are connected in the sense that the real self cannot exist independently of the formal self. Synthesis and unity are necessary for the stability of the content of consciousness. But they are also connected in such a way that the formal self can exist only to a certain degree independently of the real self. The form of the mind disintegrates whenever the content of the mind presents too many contrasts or inconsistencies.

At the beginning of his book *Den menneskelige Tanke* Høffding maintains that the mind of human beings has two essentially different modes of functioning, which are, on the one hand, "involuntary mental life" and, on the other hand, mental life as it is governed by "reflection". These two modes of functioning and the effects which arise from them do not match each other because, as Høffding says, "Involuntary mental life in its spontaneous display must not merely be described and judged in terms of how it appears to reflection. It is the sort of life to which we are not permitted to ascribe attributes and characteristics which first come into being through the work of reflection".[71] Hence, according to Høffding, the problem is that any psychological description of our mental life – governed as it must be by reflection and by linguistic means of expression, which are mostly taken from the world of reflection – can only capture *some* of the elements, and not the *whole* as it is given to us in "involuntary mental life". So the relation between reflection and "involuntary mental life" is based on an "antinomy". This antinomy expresses what might be called a relation of incompatibility between a rational understanding of psychological phenomena and our immediate experience of the same phenomena.

Already many years earlier, in the 1880s, Høffding had concerned himself not only with the incompatibility between the causal modes of description of human action and an agent's experience of acting freely, but also with the peculiar reciprocal character of psychological experiences and with the relation between conscious intellectual life on the one hand and physiological processes on the other.

Høffding claimed that rational understanding of shifting states of the mind is indeed achieved through the use of the concept of causality, in the same way as in our grasp of the physical world. Even our understanding of free will with respect to motives is possible only if the notion of causality can be extended to the domain of psychology. Høffding pointed out that efforts to understand would be useless if one and the same motive under the very same conditions at one time were succeeded by one decision and at another time by a different one. What cannot be caused cannot be the result of deliberation. Thus, he remarked, I can only command my future will in case my present will is causally connected with my future will. But he also conceived of the real self as constituting the essential motive behind actions. Motives, the moving forces of the will, are parts of ourselves, either of the real self or of the more peripheral part of it, since it depends on the nature of our personality whether a thought will be considered as a motive by us.

In spite of his determinism, Høffding also added the following in his book on Ethics:

> I cannot study the psychology of the will at the same time as taking action; but although I cannot perform these two different things at the same time, they need not be logically in conflict with each other. I cannot at one and the same time stand on my feet and stand on my head; but I can (perhaps) do first the one thing and then the other later on.[72]

How should this statement be interpreted? Høffding is not very explicit on this point in the present context. But if we put it together with other statements made by him it is possible to express what he meant.

At first glance the dichotomy between taking action and describing the causes of the free will is due to a contingent difference between theory and praxis. However, when all is said and done, this difference is based on the principal psychological distinction which Høffding termed "the antinomy between experience and reflection". According to this antinomy, involuntary mental processes can be described only insofar as we are able to grasp features of them that can be recognized as being analogous to features familiar to us from the unambiguous world of reflection.[73] The act of the will is an involuntary process, yet one of which the individual is sharply aware. Thus, being part of the involuntary mental life the act of the will is characterized as an immediately given whole, a synthesis formed by the formal self, or as Høffding also called it, by a psychic energy. The formal self, or this concentration of that energy, together with the content of the formal self forms the act of will of the real self, whence this self attains to the autonomy of a whole by constituting a real unity.

Høffding did not, apparently, distinguish between having a free will and being under the impression of having a free will because as a compatibilist who asserted that will itself is causally determined he was bound to propound, from an ethical viewpoint, the premise that the will is free, provided that we have the impression that it is free. Høffding's point then seems to have been that even though the will is causally determined, it is so only in a psychological sense. The psychological explanation of an action will specify precisely the causes of the action, and thus also the reasons for its being chosen. The process of ethical evaluation, on the other hand, finds its ultimate source in the will as the cause of the action, because the individual as an autonomous agent is immediately experienced, by himself or others, as being free to choose his action if it is consciously willed. Ethically we are interested in what precedes the will in so far as it has an influence on the will. However, ethical evaluation would not only be psychologically impossible if human actions were without causes but, further, inappropriate and unjustifiable. There is nevertheless a permanent conflict here between continuity and discontinuity, between, on the one hand, a causal description of human actions and, on the other, a description in which these actions spring from individuals taken as wholes making more or less spontaneous decisions according to the will. So the dichotomy between ethical and psychological description seems to arise from the psychological fact that in the involuntary mental life of a person he experiences himself as an autonomous being, acting through the exercise of his will, whereas on reflection he will always discover the motives behind his evaluations and actions.

The antinomy between the immediate psychological state in which concentration prevails and the mediate psychological state in which reflection is effective is revealed not only in the relation of incompatibility between actually taking action and actually studying the motives behind these actions: it exists everywhere in the psychological sphere where concentration and reflection counteract each other. Surely the reason is that a subject, concentrating on a certain mental state, experiences that state immediately as a part of itself, the subject, and only through the work of reflection is the subject able to objectify the mental state so that it becomes an object and therefore describable. Through this, the mental state is changed. Yet the two ways of experiencing the mental state supplement each other.

This situation becomes especially obvious in important areas of self-observation, where it is often impossible simultaneously to observe a conscious state in oneself and also to sustain that same state. After having discussed the possibility of dividing the ego into a part that observes and a part that is observed, Høffding continued:

> But it so happens that some of the most peculiar and important mental phenomena, such as absorption in thought, keen sense-observation, admiration, love, fear, etc., are characterized by the complete abandonment that makes such a division impossible. By being too biased in favor of this division, psychology as a science will weaken the energy of its own objects of study and yet be unable to perceive them in their natural state. Furthermore, if this observation is simultaneous with the phenomenon of consciousness, illusions will easily arise. In

haste one can easily read something into a situation that does not actually exist, or emphasize some elements at the expense of others. *Attention in itself changes the state to which it is directed. This is further accentuated by the fact that the observing and the observed parts of consciousness cannot in reality be kept entirely apart.* The expectation of finding certain thoughts or feelings in ourselves may, without our noticing anything, cause the state to be changed in the expected direction.[74]

What Høffding in fact meant can easily be illustrated by saying, for example, that one cannot feel anger towards a certain person and at the same time observe and study this emotion in its full intensity. These situations exclude each other psychologically.

The fact that a mental state is altered when the "self" whose state it is attempts to consider it objectively is a manifestation of the knowing subject's direct effect on the object that it wishes to observe. It is an "uncontrollable" effect, such that one cannot "determine" the effect and compensate for the effect so as to calculate the state of the object in the absence of the observer. This raises certain problems of description which are also encountered in experimental psychology.

It is of the greatest importance when interpreting and applying the results of experimental psychology that attention is paid to the *specific rules and conditions under which the experiments are conducted.* The conclusions that can be drawn from the experiments can naturally only be at all applicable to conscious mental life if these rules and conditions are ever present, and one must find out if this is the case before generalizing the results. It is obvious that just as we are easily able to alter our own states of consciousness by paying attention to them, it is similarly easy to alter the conscious state of a person on whom we conduct a psychological experiment, particularly when he knows that he is the object of an experiment.[75]

Once more we see the conflict between synthesis and analysis which Høffding believed manifests itself at every level of the consciousness.

Nevertheless, such complementary phenomena with mutually exclusive descriptions are not solely to be found in conscious mental life itself. According to Høffding, we find ourselves in the same situation the moment that we seek to provide a satisfactory solution to the psycho-physical problem.[76] Conscious mental life distinguishes itself from material phenomena, in Høffding's pre-quantum mechanical opinion, by displaying discontinuities.

Unconscious intervals occur in between our conscious states – if we faint, in dreamless sleep (provided that there is such a thing), and there are qualitative differences between successive states of consciousness and elements of consciousness, so that each individual element seems to be created from nothing, when we keep to strictly psychological considerations. Between different individual minds there is, at any rate, a remarkably large discontinuity. One mind can even less be derived from another than one state of consciousness can be derived from another.[77]

In addition, physiological states all have spatial extension and location, while psychological phenomena do not. The question is, then, how can physiological states have psychological characteristics? How can qualitative differences correspond to quantitative ones? How can discontinuous phenomena be linked to continuous processes?

Far from losing any interest in these questions by reason of the reduction of psychology to physiology, Høffding asserted that their intrinsic interest was enhanced by this reduction and brought out the more clearly by it.[78] But it is impossible to derive the existence of one of the series of states from the existence of the other series. We cannot reduce psychological phenomena to physiological states; but physiological and psychological reflection can supplement each other.[79] We cannot reduce mental states to bodily states, nor can we conclude that mind and body are two different entities or substances. Their mutual interrelatedness prevents their constituting two discrete entities. They are thus closely united.[80]

In his theory of the relation between mind and body, Høffding subscribed to what he called the "identity hypothesis". However, in modern philosophy of mind this term normally refers to the position which claims that mind and matter are identical in the sense that mental states are reducible to brain states. It would therefore be more correct to characterize Høffding's view as a double aspect theory, since Høffding would deny both the materialistic thesis and dualism. He would claim the truth of (1) the hypothesis that mental states and brain states are ontologically the same; but at the same time deny the truth of (2) the position that mental states can be reduced to brain states, or *vice versa*, where the materialist will accept both, and the dualist will reject both.

Høffding had rejected dualism at an early stage in his thinking on the grounds that we have no observational, and hence epistemic, access to a causal connection between brain states and mental states, and we shall never be able to have such access. Thus it is impossible to ascribe a causal connection to the relationship between mind and body. Furthermore, the assertion of a causal relation between mind and body would also contradict the thesis of the conservation of energy. All nerve processes that are converted into mental activity would entail the loss of a certain amount of energy, without this being replaced by a corresponding amount of energy.[81]

In contrast to the dualist, he claims that interaction takes place between elements that are both material and (actually or potentially) psychical.[82] Mind and body are thus two aspects of the same phenomenon. "The feelings that I have at this moment correspond to the state of my brain at this moment, *because it is the same essence that functions in consciousness and in the brain*".[83] In this situation I cannot observe and describe the physiological states in the brain without simultaneously having to forgo the description of the psychological experience which corresponds to it. And, conversely, I cannot describe my psychological experience and at the same time describe the corresponding physiological state. These two descriptions are what we could call complementary. "Mind and body are like two languages", writes Høffding, "in which reality speaks to us. We can perhaps translate from the one language to the other; but we cannot derive the one language from the other".[84] For this reason we cannot make do with one of the languages, because a complete and exhaustive description of the relation between the neuro-physiological states in the brain and states of consciousness can only be achieved through the use of both languages.

Physiology and psychology treat the same subjects considered from two different viewpoints, and there can no more be a conflict between them than between he who considers the convex and he who considers the concave side of an arc of a circle. Any phenomenon of consciousness gives rise to a double investigation. Now it is the psychical aspect, now the physical aspect of the phenomenon that is most easily accessible to us; but this in no way detracts from the fundamental relation between the two aspects, which is a mathematical function relation.[85]

Psychology and physiology are not independently exhaustive as modes of description, but together they are so. Thus they complement each other.[86]

Høffding is at pains to warn against confusing the relation between mind and body with the relation between subject and object, so as to avoid the well-known difficulties which appear in the work of Spinoza. The relation between the subject's idea of a circle and the circle existing in nature is not the same as the relation between the subject's idea of the circle and the brain process that accompanies this process of consciousness in the subject.[87] The distinction between mind and body concerns the very content of our cognition and, consequently, one can well imagine the relation to have been different from what it is, whereas no consideration of any particular content of our cognition is intrinsic to the distinction between subject and object. Nevertheless, the question is – whether or not Høffding admits it – that although there is this difference in the relation between mind and matter and the relation between subject and object, the discontinuity in the content of our cognition, given in the reciprocal relation between mind and matter, appears because of the incompatibility relation between subject and object. For, as Høffding says in the note in his *Psykologi* where he discusses this point, "Both the mental and the physical are objects for us, but while mental objects are intimately related to the knowing subject, the physical exists for us *only* as object".[88] It should thus be possible to justify the putative incompatible relationship between mind and body – if it is to agree with Høffding's other statements – by the circumstance that, under normal objective conditions, i.e., in those cases where the knowing subject *can* use the concept of causality on matters existing for consciousness, this reciprocal relation between mind and body will not be felt. But in such cases where the subject is excluded from having knowledge of any causal relation between mind and body, because part of the object is not essentially different from the subject itself, there will arise a corresponding incompatibility, as found in analogous situations in the psychology of consciousness, where either the element of concentration saturates the whole object or reflection weakens its intensity.

4. HØFFDING ON BIOLOGY

Biology represented for Høffding another field of science where the object is experienced as a whole. He claimed that the existence of this feature implies a conflict between two different modes of description. We have also heard that Høffding's teacher Rasmus Nielsen held a view according to which both mechanical and teleological descriptions of a living organism were indispensable. No doubt Høffding must have read Nielsen's book, in which he defends this position, when it was published in 1880, and he must have agreed with his view. Høffding's own position on these matters may have been influenced by Nielsen's, but the basic idea of operating with two contrasting sets of description goes back to Kant. While Høffding was still a student he had participated in a "privatissimum" which Brøchner held on Kant's *Kritik der Urtheilskraft* (Critique of Judgement) in which we find a view of biology bearing certain parallels to that which Høffding embraced later on. In January 1898 Høffding gave a lecture "Om Vitalisme" (On Vitalism) in the Biology Association in which he applied his theory of knowledge to living organisms. The idea is that the dichotomy between a teleological and a mechanical description is a consequence of the fact that the criterion of reality cannot be applied successfully to biological organisms looked upon as wholes.

At the start of his talk he distinguished between two different camps in biology: on the one hand, those who think that the mechanical viewpoint, and it alone, applies not only to physics and chemistry but to every phenomenon in nature, including living beings, and, on the other, those who deny that the organism is merely a product of organs and cells and that cells are merely products of molecules but insist that there is in living organisms a *sui generis* force acting differently from mechanical forces. These two positions he called mechanism and vitalism, following customary nomenclature. Høffding then gave a short historical survey of these two schools and the rivalry between them from antiquity onwards.

Høffding closed his lecture by throwing into relief the analogy between, on the one hand, the problem of what makes biology something more than merely physics and chemistry and, on the other hand, a whole series of problems in psychology which are all due to the conflict between the rational and the empirical requirements of a description. The factor that makes biology something different from and more than applied physics and chemistry depends on the contrast between the rational and the empirical element for our cognition, discussed on p. 75, i.e., on the antinomy between the formal requirements for the greatest possible connection between our ideas and observations that do not directly conform to such requirements. Life is precisely such a factor that cannot be explained or defined on the basis of chemical and physical concepts. Høffding expressed the matter in the following way:

> The more rational and formal our knowledge is, the greater its advance via definitions and deductions, in such a way that all transitions are made with intuitable necessity. Empirical and factual knowledge, on the other hand, must often halt at facts that could indeed be described and analyzed but not defined and derived from other facts. One such fact is life, and thus biology belongs to, and will probably always belong to, empirical knowledge.[89]

So Life is an empirical fact which at least temporarily is unresolvable into more simple components. Hence, Høffding believed that the scientific method created in biology problems similar to those in psychology. In both fields we try to integrate all phenomena into a rational matrix through definitions and deductions. But in empirical studies such a thought-system has to be verified by observations, and this is done to ascertain whether or not the relevant phenomena can be causally connected. There will, nevertheless, always be qualitative differences which must be regarded as empirical facts but which cannot be deduced from a physical base. Only to the extent to which we are able to causally connect a phenomenon with some other phenomena do we have a scientific explanation of the phenomenon in question. The fact of life is, on the other hand, an empirical fact which cannot be given a physical and chemical explanation because it doesn't fit into such a formal description.

Every time the criterion of reality fails to apply to a phenomenon we have a problem of description. This happens whenever we are faced by an individual whole which has qualitative properties other than those of the elements of which it is constituted. But we deceive ourselves, Høffding argued, if we believe that the problem can be solved by assuming new independent and original "forces", as vitalism does. The limits we encounter set up new tasks for cognition. It is these knots which challenge our faculties. Their existence generates a perpetual interaction between experience and thoughts, between induction and deduction, or between analysis and synthesis. According to Høffding, Goethe characterized this doubleness of the scientific method as the relation between inspiration and expiration. He also thought that this image portrays well the conflict between mechanism and vitalism. It is properly understood as a combat between two tendencies, between which there really is no contradiction.

If one looks closer at what Høffding was saying, it seems difficult from the context to tell whether he thought it was *logically* impossible to give a definition of life in physical and chemical terms or merely impossible *for the time being*. Sometimes he expressed himself in such a way as to suggest the impossibility of such a definition on purely logical grounds and other times as to suggest temporal grounds. However, Høffding's wavering can be seen as an expression of two features or tendencies in his philosophy. On the one hand, he believed in the non-finality of scientific knowledge, so the lack of a physical explanation of life would from this point of view be a permanent challenge for further physical and chemical investigations into its secrets. Once old problems have been solved on a purely physical basis new riddles will turn up that are apparently inexplicable in physical and chemical terms. On the other hand, Høffding also believed that biological organisms possess an individuality or a wholeness which could not be explained away by reducing this totality to the sum of all its elements, owing to the qualitative differences between the organism as a whole and its parts. Høffding first fully seemed to understand how these two tendencies could be brought into conformity with each other during his work with epistemology in the first decade of the present century.

Later, in 1925, Høffding returned to the problem of description in biology.

> From the aspect of epistemology, in my opinion, the matter is such that no limit can be set once and for all to the use of physical and chemical points of view in the organic field, but then such an application cannot in fact be carried out. The problem of the relation of an organic whole to its conditions, to the individual processes, on whose collaboration the life of the whole exists, can neither be solved by proclaiming mechanism as the solution to all riddles, nor by letting the whole "itself" intervene as a deus ex machina.[90]

The concept of "holism" cannot be used to give a scientific explanation of whatever holds the individual parts together, because "that would be to explain *idem per idem*".[91] (Therefore it is more correct to characterize Høffding's view as emergentism rather than holism: for he held that teleological concepts are indefinable in terms of mechanical concepts, and rejected the idea, most essential to holism, that the behavior of the parts can be explained in terms of features of the whole.) However, because of the unity and co-ordination life, it is necessary to take account of an emergentist viewpoint for an understanding of the relation between the individual processes and the life of the organism as such. The view of emergence is evidently considered by Høffding as a regulative principle, since no explanation of the nature of living organisms could be given by means of the concept of causality. Hence Høffding wrote, two years before Bohr introduced the idea of complementarity in quantum mechanics:

> The viewpoint of the whole is thus not in absolute contradiction to the mechanical conception. They stand in relation to each other as synthesis to analysis. And the one cannot exclude the other. But the viewpoint of the whole cannot in itself be that which yields an explanation.[92]

This was taken to mean that the tension between an emergentist account and a reductionist account will always give rise to further physical and chemical analysis, as every synthesis will make a new analysis possible. Therefore Høffding felt it necessary, on heuristic grounds, to establish holistic or teleological considerations in addition to purely mechanical considerations for living organisms, simply because there are qualitative differences between the whole and the individual constituents of an organism that cannot be reduced to quantitative differences.

I have shown that Høffding's position concerning the application of teleological concepts on biological organisms was in certain respects similar to Kant's. In *Kritik der Urteilskraft* Kant wrote that a thing exists as one of nature's purposes when it is cause and effect of itself; that is, the thing is not what it is because of its relation to something else. A tree, for example, is, according to Kant, its own cause and effect in a double sense. First, when it grows it receives and organizes matter in such a way that the whole process may be regarded as one of self-organization. Second, the leaves are produced by the tree and they are instrumental to its survival. Thus the tree can be seen as a system of mutual interdependence between the parts and the whole. Therefore we cannot conceive how organisms could have been produced by mechanical causes alone, since these operate in a blind and undesigned way. In virtue of its internal structure an organism is seen as having a natural purpose. The tree is both a self-organizing and self-productive organism in which everything is reciprocally ends and means. In fact, the principle of mutual interdependency is derived from the experience of observing organisms as wholes, but the foundation of the principle cannot rest entirely on empirical grounds. It must be based on an *a priori* principle, the idea of the purposiveness of nature. This transcendental idea takes us beyond mere sense-perception when we are using it for interpreting our experience.

However, according to Kant, such an idea of pure reason could be used regulatively only in contrast to the constitutive categories of the understanding, of which one of them, causality, is determinative of what is real. The above principle is thus regarded as a maxim for the employment of the regulative idea of judging the inner purposiveness of organisms. Hence, there is really no contradiction between a causal description of nature and claims concerning the purposiveness of beings. Although the categorical principles are constitutive of nature, and they guarantee that nature is a mechanical system as conceived in Newtonian physics, we have not proved, Kant claimed, that the description of an organism in mechanical terms is the only one possible. All we have shown is that we must attempt to press mechanical explanations of nature as far as possible because this is the only way we acquire true knowledge of nature. However, this does not exclude an alternative way of describing organisms as ruled by final causes, whenever the proper occasion presents itself. The idea of a purpose in nature, so far as natural history is concerned, is a useful, in fact, indispensable heuristic principle. But it cannot be a constitutive principle stating that the production of organisms is not possible according to merely mechanical laws. On the other hand, Kant asserted that we can approach organisms as if they were teleological systems in order to gain a certain insight into nature. However, since teleological concepts can never be given a theoretical justification and they have no objective status, we can never explain the function of organisms in this way.

Although Høffding's view can be characterized as similar to Kant's, they differ in one very essential aspect. In contrast to Kant and faithful to his empiricism, Høffding denied the existence of any necessary, *a priori* principle. He agreed with Kant that organisms are experienced as wholes and that this experience forces upon us a teleological account of biological phenomena, which has no explanatory force, and therefore is not a competitor to the mechanicistic account. Nevertheless, he insisted that the notion of a natural purpose is derived from experience itself.

5. ANALOGY AND SCIENTIFIC PROGRESS

Around the turn of the century Høffding had reached the period of his life at which his chief preoccupation was the epistemological and methodological problems of science. In 1904 he published some lectures on the modern philosophers, including, among others, Clerk Maxwell, Ernst Mach, Heinrich Hertz and Wilhelm Ostwald, lectures he had delivered at the University in 1902.[93] Through his studies of Maxwell's *The Scientific Papers* (1890) and Mach's *Die Ähnlichkeit und die Analogie als Leitmotiv der Forschung* (1902) he had acquired insight into the concept of analogy as a principle of scientific research; insight which was different from that of Aristotle, and distinct from Kant's "analogies of experience". In the same year in which he published these lectures, he stayed, on his way back from the United States, in England, where he gave a talk on 26th November 1904 in the Jowett Society in Oxford entitled "On Analogy and Its Philosophical Importance". Here he put forward the reasons why he believed that the concept of analogy is as important in science as in everyday life. First he argued that the movement of thought normally progresses through the comparisons made between different areas of experience so that one area illuminates another. He then added:

> All our cognition, spontaneous as well as scientific, is therefore full of analogies. When thinking proceeds to a new task, it does not avail itself of novel ways and means, but it tries so far as possible to make use of those which it has already applied, especially if they are clear and distinct. Language has not produced peculiar expressions for mental phenomena, but has transferred expressions which were originally conceived for material phenomena. As there are important differences between the various domains of experience, the facts not being homogeneous but constituting divers groups, each one with its peculiarities, our thinking must elucidate one group or domain by means of another, so that, in particular, the experiences which are such that they are accessible for thought in the simplest and richest way, are made use of for the understanding of others. This would not be necessary, if reality did not manifest qualitative differences. But the parts of reality, as they are known through experience, are not homogeneous, and analogy is therefore a necessary means to achieve understanding.[94]

So there are two reasons, according to Høffding, why the concept of analogy is a fundamental methodological category. Firstly, because of the existence of qualitative differences in the real world, it is impossible for cognition to apply the concept of identity to matters of fact which exhibit such differences. Nevertheless, identity is an ideal form of cognition which thought attempts to use wherever possible. Secondly, by using analogies we are able to deploy the understanding we have in one field in another which would otherwise have been unintelligible. Høffding also directs our attention to the fact that we use time-honored and meritorious concepts as far as possible to be able to grasp new domains of phenomena, and that, accordingly, the language used within one domain of experience must inevitably borrow expressions from the language used to describe a quite different domain of experience.

Analogy is defined preliminarily as "similarity of relations", that is, it consists in a comparison of similarities between the relations in one case and the relations in another case. Both in *Moderne Filosofer* and later in *Den menneskelige Tanke* Høffding explicitly subscribes to Maxwell's position on the concept of analogy by giving in Danish his translation of some lines from Maxwell's paper "On Faraday's Lines of Force", published in the volume entitled *The Scientific Papers*. Maxwell wrote, "By a physical analogy I mean that partial similarity between the laws of one science and those of another which makes each of them illustrate the other. Thus all the mathematical sciences are founded on relations between physical laws and laws of numbers, so that the aim of exact science is to reduce the problems of nature to the determination of quantities by operation with numbers".[95] However, although Høffding's Danish translation did not quite accurately render the meaning of the original, it was, in a sense, even more interesting with respect to Bohr's ideas on correspondence. As part of the Danish translation has it, "By a physical analogy I mean the partial similarity between the laws of two domains of experience which brings it about that one illustrates the other". Maxwell had emphasized the symmetrical nature of an analogy, while Høffding laid stress in his translation on its asymmetrical nature. At all events, Høffding construed Maxwell's statements as saying that the application of the concept of numbers to real items is based on an analogy between the series of numbers and the series of items.

As mentioned earlier, analogy is tentatively defined as similarity of relations between two items or phenomena. Such similarities are not concerned with the individual properties of these items, nor with single parts of them, but bear mutually upon the relation between the properties or the parts. For Høffding analogies fell into two main types: The quantitative analogy is the same as "identity of relation", an analogy which may also be called proportion. An example of such an analogy is the relation between two numbers being identical with the relation between two other numbers. The other type, the qualitative analogy, can be defined as an "identity of difference", since any difference is a qualitative difference, and may exist, for instance, between geometrical figures or between the organs of different organisms. Furthermore, Høffding lays down the conditions which scientific analogies must satisfy in order to be considered sound:

> They must not only fit on single points, but on all points, so that full consequences may be drawn from them and be applied in the explanation of details. The symbol or the image must serve us as a schema which we may follow accurately in special cases, and which can lead us to new experiences.[96]

However, a qualitative analogy may also be applied in the case of more complex systems of relations, such as those existing between a quantitative and a qualitative series. In the series of quantity and those of quality there is an identity of difference between the members of the series. It was Høffding's belief that an essential aim of natural science is to establish such an analogy between a qualitative series and a quantitative series.

In Høffding's view it is in the face of breaks in what we experience as the feature of connectedness that the use of analogies is appropriate: they serve to remedy what would otherwise constitute an incoherence in the description of our experience. The feature of connectedness, which invariably, but involuntarily, informs the act of cognition, and is indeed required by the very nature of the act, is that which regulates the process of acquisition of knowledge. When this feature is not supplied in cognitively informed experience, as in cases where we are confronted with the unfamiliar, we resort to analogies in our attempt to re-establish coherence. The importance of analogy for us appears, Høffding says, in three ways: Either the analogy directs further research in the sense that it guides our cognition in the exploration of unknown phenomena; it so to speak aids the cognitive grasp of the unknown phenomena through a comparison with known phenomena. Or it may unite different items or phenomena which cannot be reduced one to the other but which can be connected with respect to the similarities involved in the analogy. In the first case the analogy bridges the gap between our past and present knowledge, on the one hand, and our future knowledge, on the other. In the second case it gives us an insight into areas between which no continuity exists. Finally, there is the type of analogy which is expressive of poetic and religious feelings and which is not based on anything that can be demonstrated in experience.

Analogy is a necessary and a justified methodological instrument of scientific thought, according to Høffding. The history of science has shown how one domain of experience may constantly be used in order to elucidate another domain, and how the fruitfulness and the tenability of the working hypothesis involved in analogy is tested by experience.

In 1922, when Høffding was writing his essay on analogy, a concept which he had always regarded as an indispensable one but had placed second among the formal categories of human thinking after the concept of identity in "Totalitet som Kategori" (1917), he had an exchange of letters with Bohr about the subject:[97]

Carlsberg, 20th Sept., 1922

Dear Professor Bohr,

As I mentioned to you in the summer I would like to put you a question with reference to your essay on "Atomernes Bygning og Stoffernes fysiske og kemiske Egenskaber".[98]

I have noticed that you often use expressions by which you indicate a relation of *analogy* (not a relation of identity) between the structure of the atoms and the actually present physical and chemical data. Such expressions as "illustration" (p. 1), – "explanation or rather understanding" (p. 33), – "interpret" (p. 36), "as the spectrum teaches us and as the atomic model makes intelligible" (p. 45).

My question is whether the term "analogy" would not comprise the sense of the expressions used by you at various points about the relevant issues. All understanding is based – except that of pure logic – on analogy, and science is a strictly rational systematization of analogies between different domains of cognition. Thus, according to Hjelmslev (and according to Zeuthen in his most recent essay) there is an analogical relation between algebra and geometry; and I believe that this idea can be brought to bear upon relations in every area, including that dealing with the relation between humane studies and natural science. Thus I make a fundamental distinction between analogies that can be brought about rationally, and other analogies that appeal to poetic and religious feelings, and which cannot be maintained coherently.

Naturally, I don't want to draw you into a philosophical discussion, but I would like to know your opinion, before I make use of your essay in a book on the concept of analogy on which I am working.

With kind regards,
Yours sincerely Harald Høffding

Bohr responded very quickly to the letter from Høffding by sending him a typed letter expounding his view.[99]

22th September, 22

Dear Professor Høffding,

Thank you for your very nice letter which interested me very much. The relation you wish to emphasize concerning the role of analogy in scientific investigations is without doubt an essential feature of all studies in the natural sciences, although it is not always a salient one. It is often possible to use a picture of a geometrical or of an algebraic nature which in a very clear way captures the problems in question to an extent which is adequate in the context, so that what is under consideration almost acquires a purely logical character. In general and particularly in new fields of research, one must, however, constantly have the evident or the possible in mind, and be satisfied if the analogy is merely striking enough to ensure that in so far as this picture applies its usefulness or rather fruitfulness is beyond any doubt. Such a relation applies not least to the present situation in the theory of the atom. Here we are in the peculiar situation that we have acquired certain information about the structure of the atom which can be considered as being as certain as any one of the facts in the natural sciences. On the other hand we encounter difficulties of such a deeply-rooted nature that we do not even faintly see the road to their solution; in my personal opinion these difficulties are of such a nature that they hardly allow us to hope that we shall be able, in the world of the atom, to carry through a description in space and time of a kind which corresponds to our ordinary sensory images. Under these circumstances one must naturally constantly bear in mind that one is operating with analogies, and this step, in which the application of these analogies is delimited in each case, is of decisive significance for progress.

If you do wish to hear more about the two examples you referred to in your letter, or any other, I am naturally prepared to assist you to the best of my ability with any information. You will find enclosed a small leaflet giving a German version of the lecture you referred to, plus two earlier lectures, not with the intention of tiring you further with physical details but only because you will find in the introduction some remarks which, even though in a philosophical sense they lack definite form, will, however, show you how much the relation in question concerns me.

With kindest regards from my wife and myself,
Yours sincerely, Niels Bohr

In his book on analogy, it is quite clear from the passage in which Høffding discusses Niels Bohr's paper what Høffding intended.[100] He considered the physicists' work on the mathematical description of the periodic system on the basis of the observable physical and chemical properties of the elements as an example of the type of analogy between algebra and the physical laws, between a series of differences among numbers and a series of differences of qualitative properties, to which Maxwell had drawn attention. However, Høffding did not mention Bohr's notion of correspondence, to which Bohr apparently referred in the end of his letter to Høffding.

The strong emphasis Høffding laid on the concept of analogy as one of the fundamental methodological principles in science must have influenced Bohr deeply already in his student days. He never let it slip from his mind. In fact, when Bohr developed his theory of the structure of the hydrogen atom, it was on the basis of the methodological precept he had learned from Høffding. In a letter to Rutherford of 23rd March 1913 he spoke of "the most beautiful analogi [Bohr's Danish spelling] between the old electrodynamics and the considerations used in my paper".[101] So when he eventually formed the idea of correspondence it was at once given the name "a formal analogy".

PART II

Bohr and the Atomic Description of Nature

Chapter V

1. THE PRINCIPLE OF CORRESPONDENCE

Already early on the principle of correspondence had been a guide for Bohr's research on the radiation properties of atoms, with respect to which it had an exact, technical meaning. Later on, after Bohr had introduced his ideas of complementarity, it was used in a more general and broader sense, as we shall see in a subsequent chapter. The correspondence principle is a generalization of the fact that results derived from Bohr's theory of hydrogen in the limiting region of high quantum numbers coincide approximately with those yielded by classical electrodynamics. The principle states that such a coincidence must hold generally in all cases involving high quantum numbers; hence it became the methodological principle which guided Bohr's research in his endeavour to establish a coherent quantum theory during the 1910s and 1920s. However, although Bohr and others little by little were able to explain many spectroscopic data on the basis of the correspondence principle, they never made a decisive breakthrough. On the contrary, at the beginning of the twenties Bohr's theory of atomic structure was confronted with serious problems, such as that concerning the determination of the energy states of any atom other than hydrogen atoms and the anomalous Zeeman effect, problems with which it failed to cope. In the end it was not Bohr himself but Werner Heisenberg, his young assistant, who in 1925 articulated the foundation of a coherent quantum theory for which Bohr had been searching for so long. Nevertheless, he saw Heisenberg's theory as "a precise formulation of the tendencies embodied in the correspondence principle".[1]

What marked the new era in physics to which Bohr contributed so greatly in 1913 by putting forward his theory of hydrogen was Planck's discovery of the quantum of action in 1900. In order to explain the radiation spectra from black bodies Planck had to assume that the energy was quantized in the sense that there was a lower limit to the amount of energy which could be emitted with a certain wavelength from black bodies with a certain temperature. Planck himself only recognized that if he used a certain physical quantity, which he called the elementary quantum of action, then his equation would agree with the measured intensities. But what the quantum of action really denoted was not clear at that time. This first became clear with the work of Einstein, who five years later got the idea that the quantum of action had something to do with the amount of energy which was bound to the light waves. He imagined that the energy in the light waves was concentrated into small lumps – photons – whose magnitude had to be determined by the quantum of action and their wavelength, and that the energy and the wavelength of the photons had to be inversely proportional. The exact expression which Einstein found to be connecting these quantities is $E = hc/\lambda$, where E is the energy of the photon, h is the quantum of action, called Planck's constant, c is the velocity of light, and λ is the wavelength of the photon. So, for the first time, Einstein had created a conception of light as a duality consisting at the same time of waves and particles, a duality which ever since has perplexed all who have philosophized about the foundation of quantum mechanics.

The principle of correspondence was put into operation by the initial success of Bohr's theory of hydrogen, which was the next step towards a consistent quantum mechanics. This theory departs from classical electrodynamics by being founded on two non-classical presumptions. The first one is the quantizing condition stating that an electron moving around the nucleus of the atom can only exist in certain states, the so-called stationary states, where the energy of the electron has a well-defined value without the system emitting radiation. According to classical electrodynamics the energy of the electron might have taken any value while orbiting around the nucleus continuously, emitting radiation, until the time when the electron was swallowed by the nucleus. The second presumption is the truth of the frequency condition, which claims that in passing from one stationary state to another the electron emits radiation with a frequency which corresponds to the difference between the energy of the two stationary states divided by Planck's quantum of action. In classical electrodynamics it is assumed that the frequency of such radiation depends entirely on the mechanical frequency of the electron during its approach towards the nucleus and not on the energy value of the electron after it has completed a transition. Taken together these two non-classical conditions yield $h\nu = E_n - E_m$, relating the energy $h\nu$ of the emitted radiation to the energy difference $E_n - E_m$ between the two stationary states involved in the transition. So according to Bohr's theory, as the energy difference converges the frequency ν tends towards zero. This is the case in the limiting region of high quantum numbers where the stationary states are far from the nucleus. In that region the frequency of the emitted radiation coincides asymptotically with the frequency of the revolution of the electron and the intensity of the emitted radiation coincides approximately to the amplitude of the harmonic component of the periodic motion of the electron.

In the important work written in 1918, "On the Quantum Theory of Line-Spectra. Part I", in which Bohr discussed the possibility of using quantum theory for determining the line spectrum of an atomic system without the necessity of introducing assumptions concerning the mechanism in the transition between two stationary states, the principle of correspondence is formulated as follows: it can be shown that the frequencies, calculated on the basis of the law of frequency,

> ... in the limit where the motions in successive stationary states comparatively differ very little from each other, will tend to coincide with the frequencies to be expected on the ordinary theory of radiation from the motion of the system in the stationary states.[2]

At this stage the principle had not been given the name "principle of correspondence" but was termed a "formal analogy between the quantum theory and the classical theory". This formal analogy was the methodological principle behind Bohr's research into the structure of the atom. On pages 15–16 Bohr gives a more exact explanation of the nature of the analogy for a system of one degree of freedom:

> In order to obtain the necessary connection, mentioned in the former section, to the ordinary theory of radiation in the limit of slow vibrations, we must further claim that a relation, as that just proved for the frequencies, will, in the limit of large n, hold also for the intensities of the different lines in the spectrum. Since now on ordinary electrodynamics the intensities of the radiations corresponding to different values of τ are directly determined from the coefficients C_τ in (14) $<\xi = \Sigma\, C_\tau \cos 2\pi\, (\tau\omega + c_\tau)>$, we must therefore expect that for large values of n these coefficients will on the quantum theory determine the *probability of spontaneous transition* from a given stationary state for which $n = n'$ to a neighboring state for which $n = n'' = n' - \tau$. Now, this connection between the amplitudes of the different harmonic vibrations into which the notion can be resolved, characterized by different values of τ, and the probability of transition from a given stationary state to the different neighboring stationary states, characterized by different values of $n' - n''$, may clearly be expected to be of a general nature. Although, of course, we cannot without a detailed theory of the mechanism of transition obtain an exact calculation of the latter probabilities, unless n is large, we may expect that also for small values of n the amplitude of the harmonic vibrations corresponding to a given value of τ will in some way give a measure for the probability of a transition between two states for which $n' - n''$ is equal to τ. Thus in general there will be a certain probability of an atomic system in a stationary state to pass spontaneously to any other state of smaller energy, but if for all motions of a given system the coefficients C in (14) are zero for certain values of τ, we are led to expect that no transition will be possible for which $n' - n''$ is equal to one of these values.

From 1918 to 1920 the correspondence principle was known and referred to as Bohr's principle of analogy between the quantum theory and classical electrodynamics. This appears from a letter from A. Sommerfeld to Bohr on 5th February 1919.[3] This use of "analogy" is interesting, because, as we saw earlier, Høffding often used the term "analogy" in cases where our cognition cannot make the content of cognition fully conform to the identity relation that is the ideal of epistemology. By pointing to this analogy, even though radiation phenomena cannot be explained on the basis of classical electrodynamics, Bohr ensured that quantum theory retained a modicum of rationality.

The first occasion on which Bohr used the words "correspondence principle" was in "On the Series Spectra of Elements" from 1920. Here he says:

> Although the process of radiation can not be described on the basis of the ordinary theory of electrodynamics, according to which the nature of the radiation emitted by an atom is directly related to the harmonic components occurring in the motion of the system, there is

found, nevertheless, to exist a far-reaching *correspondence* between the various types of possible transitions between the stationary states on the one hand and the various harmonic components of the motion on the other hand. This correspondence is of such a nature, that the present theory of spectra is in a certain sense to be regarded as a rational generalization of the ordinary theory of radiation.[4]

In 1922 Bohr explained in an Appendix to "On the Quantum Theory of Line-Spectra. Part III" why he had changed the terminology:

Note to § 1. The problem treated in this paragraph offers a simple application of the point of view developed in Part I and denoted there as a formal connection, or analogy, between the quantum theory and the classical electromagnetic theory of radiation. In order to prevent the possible misunderstanding that it is here a question of a direct connection between the description of the phenomena according to the quantum theory and according to classical electrodynamics, in later papers of the author the law in which this analogy appears is designated as the "correspondence principle".[5]

However, the last sentence indicates quite clearly that Bohr still regarded the principle of correspondence as a principle of analogy in Høffding's sense.

Drawing upon Bohr's written and spoken words we may give a general formulation of the principle of correspondence. In its original sense it refers to a formal connection consisting in a structural similarity between two distinct theories created by a rational generalization of the older theory in making the new one. The formal analogy between classical electrodynamics and quantum mechanics consists in the fact that with each of the quantum transitions there can be associated a certain Fourier component of the classical motion of the electron, and that for large quantum numbers the radiation emitted coincides with that emitted classically by the corresponding Fourier component, as well in the fact that the probability of transition can be associated with the square of the Fourier component of the dipole moment with which it will agree in the limit.

Philosophers of science have construed it as if correspondence in Bohr's sense were a kind of formal reduction, that is, a theory T_2 may be said to correspond to another theory T_1 if and only if some or all the law statements of T_2 can be reduced to law statements of T_1 under certain constraints on the constants of T_2.[6] Indeed, this does not mean that the laws of T_2 become identical or can be transformed into those of T_1 since the constants do not disappear under such a formal reduction. The mathematical formulation of the law statements of the two theories are distinct; it is only the numerical solutions yielded by the theories which are practically identical when the magnitude of Planck's constant h is negligible. Thus, quantum mechanics reduces to classical mechanics if the constraint on h is that it is zero. But without such a restriction the two theories are logically inconsistent.

That Bohr around 1920 had something like this in mind seems very possible. The principle of correspondence involves in other words a similarity of structure between the original and the new theory so that the latter with certain formal restrictions reduces to the former. But must correspondence be understood as a formal and syntactical principle? Is the rational generalization Bohr is talking about purely formal? Indeed not. Even though Bohr spoke of correspondence, at that time, as a formal connection or a rational generalization, the principle must also presuppose or imply certain semantical requirements according to which the meaning of the observables common to both theories is empirically the same or almost the same. The original theory, whose domain of application has been narrowed due to the incidence of anomalies outside its proper limit, is still supported by all the experience available within its new restricted domain of application. Thus the new theory has to have a form as well as a content that is supported by the very same experience as well. It is a requirement made on a new theory that the domain of its application has to include that of the theory it supersedes, so that the new theory will be capable of explaining all the hitherto successfully explained experience. Consequently, it is a requirement that many observational quantities of the new theory must be identical with those of the old for both theories to make approximately the same predictions with respect to the domain of experience common to both. In fact the predictions of classical mechanics and quantum mechanics can only coincide empirically, in spite of their conceptual differences, if any observational expression within quantum mechanics (under certain constraints) is reducible to an equivalent observational expression of classical mechanics. The reduction is possible only if all the observational quantities referred to in the expressions of classical mechanics are identical to or identifiable with some combination of those referred to in the expressions of quantum mechanics. Such a requirement should guarantee that the laws which have proved successful inside the restricted domain of the superseded theory also prove successful inside the domain of new theory by letting the predictions of the new theory at its limit coincide with the predictions of the older theory in such a way that the quantitative differences between the two predictions become empirically negligible or empirically undetectable.

A semantic approach to the correspondence of theories would involve the concept of a model. Briefly, a model of a theory is any interpretation under which the axioms of a theory are true or satisfied, and an interpretation consists of a domain of entities over which the individual variables and the predicate variables of the theory range. So we may roughly say that a theory T_2 having a domain of application D_2 corresponds with a theory T_1 having a domain of application D_1 if and only if (1) D_1 and D_2 have a common domain D_c, and (2) every model M_2 of T_2 is isomorphic to a model M_1 of T_1 within D_c but M_2 of T_2 is not isomorphic to M_1 of T_1 outside D_c. This means that there exists a correspondence between T_2 and T_1 if all the interpretations of T_1 are isomorphic to the interpretations of T_2 within D_c in such a way that T_1 and T_2 differ empirically with respect to their domains outside D_c but are empirically equivalent with respect to D_c.

The conclusion we must draw then is that the principle of correspondence cannot merely be characterized as a formal connection between two theories or as a purely syntactical generalization, as Bohr was inclined to do around 1920. Generally speaking, correspondence can only operate as a heuristic or methodological rule guiding the formulation of a new theory if it involves a syntactical as well as a semantical requirement of what the relation between the superseded theory and the new theory should look like. For although the theoretical meaning of the terms of the two theories differs, since this kind of meaning is determined by the conceptual structure of the theory, the empirical meaning of the terms of the two theories have to be similar. Thus the generation of the new theory has to take its rise from the descriptive categories of the old theory, thereby incorporating its domain of application into the domain of application of the new theory. It must do this in a way that explains why the old theory did not fail within its restricted domain, but was doomed to fail when it was considered as a general theory capable of explaining the enlarged domain of the new theory. Moreover, as Høffding pointed out, the heuristic value of using an analogy consists in the attempt at understanding the unknown on the basis of what is known, because new phenomena are intelligible in so far they can be seen to be analogous to already well-understood phenomena. This implies that the new theory has to operate with many of the same descriptive categories as the old one and that therefore the empirical content of the language in which the new theory is formulated is similar to the empirical content of the language of the old theory, a language which it is necessary for us to use in referring to already familiar experience. So by using analogies the language of the old theory is more or less imposed upon us in our effort to bring new experience into a descriptive connection with well-established experience. Thus, according to Høffding's methodology but as opposed to a Kuhnian methodology, the old and the new theory must be commensurable with respect to empirical meaning. In other words, the semantical requirements of using analogies imply that two corresponding theories have to be empirically comparable even though they may be logically incompatible. The two theories may be based on widely differing assumptions regarding certain aspects of physical reality, and hence the theories may involve different ontological commitments, but the empirical content of the language in which these assumptions are expressed is the same or similar. This analysis of the preconditions of the principle of correspondence *qua* a methodological rule is, as I take it, much more in harmony with the broader scope which Bohr gave the principle later on, after he had introduced the notion of complementarity. At that time Bohr elaborated his reasons for believing that the generation of any quantum theory always has to take its point of departure in the classical theory.

In this connection it may be noted that it is correct of Max Jammer to stress the emphasis Bohr laid on the precept that "the correspondence principle must be regarded purely as a law of the quantum theory, which can in no way diminish the contrast between the postulates and electrodynamic theory".[7] However, it was perhaps unwise of Bohr at that time to call the principle of correspondence a law of quantum mechanics. It can be seen as a heuristic principle which tells us how to construct a consistent quantum theory on the basis of classical mechanics. But it can never be part of the quantum theory itself. Indeed, Bohr indirectly acknowledged this himself when after 1927 he extended the application of the principle beyond the connection between classical mechanics and quantum mechanics by regarding the relation between the theory of relativity and classical mechanics as a manifestation of the very same principle.

Although Bohr at this time called the principle of correspondence a law of quantum theory, Max Jammer is not correct in supposing that the principle thereby conflicts according to Bohr's view with the need to use classical concepts in quantum mechanics. At least he did not feel so. For as Bohr explicitly said at the conclusion of "On the Application of the Quantum Theory to Atomic Structure":

> As frequently emphasized, these principles [of which the correspondence principle is one], although they are formulated by help of *classical conceptions*, are to be regarded purely as laws of the quantum theory, which give us, notwithstanding the formal nature of the quantum theory, a hope in the future of a consistent theory, which at the same time reproduces the characteristic features of the quantum theory, important for its applicability, and, nevertheless, can be regarded as a rational generalization of classical electrodynamics.[8]

In the introduction to the same work, Bohr put it even more explicitly:

> From the present point of view of physics, however, every description of natural processes must be based on ideas which have been introduced and defined by the classical theory. The question therefore arises, whether it is possible to present the principle of quantum theory in such a way that their application appears free from contradiction.[9]

This Bohr wrote in 1922. But of course the principle of correspondence is not a law of quantum mechanics but a methodological principle. Such a principle cannot in itself set limits on ontological models of a theory, hence it cannot be in conflict with quantum theory. However, since the correspondence principle is a heuristic rule for giving quantum physics a structure analogous to that of classical physics it imposes the retention of classical concepts in order to guarantee the same empirical content of the two theories in their common domain of application. The problem which Bohr was to confront in the succeeding years was how to create a consistent theory as well as a coherent interpretation based on classical concepts.

2. THE SEARCH FOR AN INTERPRETATION

It was not until Heisenberg's success in early summer of 1925 in formulating a formalism of quantum mechanics, the so-called matrix mechanics, that it became possible to describe all quantum phenomena. This formalism is based on the idea of representing observable quantities by sets of time-dependent complex numbers. The aim was that of describing experimental results without having to visualize atomic objects in space and time.

The classical ideal of a description presupposes the assumption that either the state of a closed system remains unaffected by observation or, if not, that it is always possible to account for the effects of observation when describing the state of the system in isolation from it. Through observation it is possible to determine the motion of a system by registering its trajectory in every point of space and at any moment of time. But when there occurs observational interaction with the system it must be possible to determine the effect of the interaction on the state of the system if it is to be possible to define the state of the isolated system subsequent to the interaction. This is only possible given the principle that momentum and energy are conserved. Thus in the classical framework two modes of description are combined: that in which the state of a system develops continuously in space and time, and that in which a change of the state of a system caused by interaction is determined by the principles of conservation of momentum and energy. From this it follows that the isolated system can always be ascribed a well-defined mechanical state irrespective of whether the system interacts with another system or not. It is in fact this combination of a causal description given in terms of energy and momentum conservation with a description with respect to every point of space and time, which in the classical framework yields the deterministic description of the system, and which allows us to define any future state of an isolated system as soon as we have determined its initial state by observation.

This may help us to understand why at one time Bohr had misgivings about Werner Heisenberg's proposal to abandon entirely any attempt at visualizing the behavior of atomic phenomena and which led him instead to postulate "That nature allowed only experimental situations to occur which could be described within the framework of the formalism of quantum mechanics".[10] Such a position would imply the renunciation of the classical ideal of description: if the precise use of concepts of space and time coupled with a precise use of the conservation theorems for energy and momentum was no longer possible, then the ideal of continuity in nature could no longer be upheld.

Heisenberg put forward his proposal to abandon any visualizable model of the atomic system in discussions with Bohr and Kramers while staying in Copenhagen during the winter of 1924–1925. In "Quantum Theory and Its Interpretation" he indicates that he had managed to convince Bohr of the necessity of giving up any visualizable picture when he returned to Copenhagen late in the summer of 1925. Up to that time Bohr had still hoped that it would be possible to describe the stationary states of the atom in classical terms in contrast to a description in terms of the transition between such states.[11] In his atomic model of 1913 Bohr had imagined that the states of the atomic systems could be represented by a classical "model" or "picture" in which the electrons have spatial loci at each temporal instant, forming a continuous orbit around the nucleus. Furthermore, until 1923 Bohr had chiefly devoted his time to the study of the quantum theory of atomic structure and he had been very little concerned with the quantum theory of radiation. He held that the emission and absorption of radiation by an atom occurred in discrete quanta of energy, although he still assumed, contrary to Einstein, that the radiated energy is transmitted in continuous wave fronts *in vacuo*. Bohr thought that Einstein's hypothesis of quantized photons should be looked upon as a heuristic proposal without any claim to being a literal representation of the propagation of light. Instead he hoped that it would eventually become possible to describe the interaction between radiation and atomic systems according to a field representation of radiation.

Owing to this reluctance to accept the photon hypothesis as implying a limit on the field representation and his inclination towards the continuous wave hypothesis, in 1924 he presented, together with Kramers and John Slater, a paper in which it was assumed that the spontaneous transitions between stationary states of an atom are induced by the virtual radiation field produced by the atom, as well as the virtual radiation produced by other atoms in such a way that the energy is not strictly conserved in the individual interaction between matter and radiation, but only conserved statistically over an average of many interactions.[12] The basic conception behind Bohr's suggestion was apparently this: since it is the application of the principle of conservation of energy that makes the description of a continuous change of state possible, then the existence of discontinuous changes in state implies that the energy is not conserved in each individual interaction but only in the long run. By abandoning the principle of energy conservation, Bohr was able to combine the discontinuously changing atomic system with the continuously changing fields. Nevertheless, about a year after this radical view had been made public it was proved that energy was conserved in every single case. After the failure of the attempt to jettison the notion of energy conservation Bohr turned towards the possible rejection of the other component of the classical framework of description: the ordinary space-time description of nature.

Max Jammer thinks, however, that in 1918 in "On the Quantum Theory of Line-Spectra" Bohr had already expressed himself in such a way as to suggest that he did not include among the assumptions of quantum mechanics the union of space-time description and the energy conservation principle of the classical framework. When Bohr said that the system "will start *spontaneously* to pass to the stationary state of smaller energy"[13], he did not then mean, as Einstein did, only the lack of excitation from external causes but also excluded in principle the possibility of the activity of "internal" parameters.[14] However, the fact that this interpretation does not seem to hold good can be seen from a related remark made by Bohr in 1922 in "On the Application of the Quantum Theory to Atomic Structure", where he explicitly draws attention to spontaneity in relation to external causes.

> Thus we shall assume, with Einstein, that an atom in a stationary state possesses a certain probability of shifting to another state of smaller energy within a given element of time, with the emission of radiation. This occurs spontaneously; that is, without any assignable external stimulation.[15]

So there is nothing in what Bohr says here which disproves the claim that, until 1921–1922, he was guided in his work on atomic structure by the view that stationary states should be described strictly mechanically, while the transition between two stationary states could not be explained by external mechanical influence. As late as 1924 Bohr hoped, in the paper written with Kramers and

Slater, to be able to explain the transition by combining a classical picture of the stationary states and virtual radiation. Nevertheless, not long after the disconfirmation of the Bohr-Kramers-Slater theory Bohr finally admitted the necessity of forgoing the possibility of any form of visualization of the atomic system in space and time; he had evidently reached the conclusion that the non-classical discontinuity which manifests itself in the transition between stationary states was the factor that had to be reckoned with in any account of the stationary states themselves as well as the interaction between the atomic system and radiation.

Heisenberg's recollection of the discussion between Bohr and himself has become the official version of Bohr's opposition around 1924–1925 to a purely mathematical and formal account of the atomic phenomena which ascribed properties only to observable phenomena resulting from the interaction between the atomic system and radiation. But at that time Bohr was not totally unfamiliar with the idea of the impossibility of visualizable models. Already in 1922, when he wrote "On the Application of the Quantum Theory to Atomic Structure", he seems to have anticipated the possibility of giving up every space-time description of the structure of the atom. In his letter to Høffding written on 22nd September 1922, quoted on page 108, Bohr emphatically expressed as his personal opinion the view that the difficulties of grasping the structure of the atom "hardly allow us to hope that we shall be able, in the world of the atom, to carry through a description in space and time of a kind which corresponds to our ordinary sensory pictures". So apparently even before Bohr attempted to retain the classical space-time description through launching the Bohr-Kramers-Slater theory, he must have oscillated between contemplating a departure from the conservation principle or giving away the classical space-time description. But once the conservation principle of momentum and energy at the microscopic level was confirmed in the spring of 1925 by the Bothe-Geiger and Compton-Simon experiments, Bohr very quickly realized that a further understanding of the atomic system could be reached only if the other component of the classical framework was brought into question. As it turned out, it was Bohr's renewed interest in the problems issuing from a complete abandonment of the hope of realizing the visualizability of the atomic system in space and time which in the end led him to complementarity.

Thus, as opposed to many other physicists, Bohr's first reaction to the refutation of the Bohr-Kramers-Slater theory was not to accept the photon hypothesis, but to put the blame on the use of spatio-temporal pictures in the domain of quantum phenomena. Bohr believed that the Bothe-Geiger results proved the impossibility of a classical spatio-temporal description of the interaction between radiation and matter rather than the correctness of the corpuscular hypothesis of light.[16] And commenting on the Bothe-Geiger and Compton-Simon experiments he wrote:

> From these results it seems to follow that, in the general problem of the quantum theory, one is faced not with a modification of the mechanical and electrodynamical theories describable in terms of the usual physical concept, but with an essential failure of the pictures in space and time on which the description of natural phenomena has hitherto been based.[17]

The reason he still regarded the corpuscular hypothesis as formal and not as describing anything real was because he found the definition and measurement of frequency contained in the expression of the energy of light quanta as resting "exclusively on the ideas of the wave theory".[18] What Bohr meant was that the classical ideal of description reflected in the mechanical and electrodynamical theories, according to which such processes could be visualizable in space and time, failed in the quantum world because of "an element of discontinuity in the description of atomic processes quite foreign to the classical theories".[19] But during 1926 Bohr seemed gradually for the first time to have recognized, probably under the influence of the publication of Schrödinger's wave mechanics during the early months of 1926, that the particle model of light was neither more nor less formal than the wave model. In a way this insight may be characterized as the final breakthrough in his understanding of the quantum world.

A decisive step on the road towards the orthodox interpretation of quantum mechanics had already been taken in the autumn of 1923 when Louis Victor de Broglie originated the idea which was the converse of Einstein's, viz. that if light waves can also be described as particles, perhaps particles may be regarded as waves, too? The problem was only that if this were the case, then Einstein's formula could not be used just as it was, because no particle moves with the velocity of light. De Broglie was, however, able to modify Einstein's formula so as to accommodate particles. At that time it was known that photons also have a momentum p besides an energy E, and that they were connected by the equation $E = pc$. So if this formula is combined with Einstein's, c is eliminated and the result becomes $p = h/\lambda$. This can only be understood as the claim that all particles have a wavelength whose value is dependent on how fast they move. Even macroscopic objects have a wavelength, though it is much smaller than the objects themselves, smaller even than the radius of the proton.

Experiments conducted in 1927 by C.J. Davisson and L.H. Germer with electrons confirmed the correctness of de Broglie's assumption. Electrons could be diffracted and interfered with just like waves. The famous double-slit experiment serves as a good example of the latter property. One by one electrons are emitted towards a screen containing two small slits. On the other side of the screen there is a photographic plate which records each electron as it moves through one of the slits. Given the acceptance of classical point of view we would now expect after the dispatch of many electrons that an assemblage of spots has been built up on the photographic film behind each slit in the screen, so distributed as to yield fewest spots at the periphery. This is in fact what happens if first one of the slits has been kept closed so the electrons have only been able to pass through the other and if, subsequently, the other slit has been closed and the first has been open. But if both are kept open, the result is different. Behind the two slits an interference pattern appears which is similar to the pattern made by a wave when it is separated into two parts and these cause interference on the other side of the screen. But this will happen only if the distance between the slits is not much greater than the de Broglie wavelength of the electrons. The particulate nature of the electron, however, is still present in that it leaves a highly localized darkening on the plate when interacting with the molecules of the photographic emulsion, but at the same time every single electron seems capable of "sensing" both slits in its path between the source and the plate and is thus able to interfere with itself. So in a single experiment electrons behave both like particles and waves, as something which is both localized and spread out into space.

Bohr's reaction to the duality between continuity and discontinuity inherent in quantum mechanics was to deny the realistic significance of the classical models of particles and waves. Contrary to what Schrödringer hoped, viz. that his wave mechanics based on de Broglie's ideas could reestablish continuity in quantum mechanics, Bohr now expressed in the autumn of 1926 in discussions with Schrödinger and Heisenberg in Copenhagen his conviction that "all the apparently visualizable pictures are really only to be regarded symbolically" without any realistic interpretation of the mathematical formalism of the wave function.[20] Towards the end of 1926 it was clear to Bohr that

> The quantum theoretical description of the atoms contains an essential element of discontinuity which stands in an unquestionable opposition to the demands of the classical mechanical and electromagnetic theories. However, the promising results which in recent times have been reached with the aid of the modification of classical theories known as wave mechanics has once again raised the question of the possibility of avoiding every element of discontinuity. At the present stage of science this possibility does not seem to exist since here it concerns difficulties in the fundamental concepts themselves on which wave mechanics as well as the classical theories rest.[21]

This statement is a translation of the abstract of Bohr's communication to the Royal Academy on 17th December 1926, but no manuscript seems to have survived. In a letter of 30th December Høffding reported to Meyerson about the communication:

> In his recent lecture on "Atomic Theory and Wave Mechanics" he suspects that we cannot decide whether the electron is a wave motion (in which case we could avoid discontinuity) or a particle (with discontinuity between the particles). Certain equations lead us to the former inference, certain others to the latter. No picture, no term corresponds to all equations. – In a conversation which he had with me after the lecture Mr. Bohr told me that he is ever more convinced of the necessity of symbolization if we wish to express the latest findings of physics. – Quite certainly he will publish his latest scientific results.[22]

Nevertheless Bohr did not publish his talk. During the following months he constantly improved his understanding of the particle-wave duality.

In fact, Heisenberg was the first person who publicly contributed to the orthodox interpretation of quantum mechanics. During the period which followed his invention of the matrix mechanics, a debate was initiated about how the physical content of the formalism was to be understood. In the early part of 1927 Heisenberg's efforts were crowned with success as he was able to derive his famous uncertainty relations on the basis of the Dirac-Jordan generalization of his own matrix mechanics. To support his interpretation of the uncertainty relations he analysed the situation which followed from the particle-wave nature of matter in connection with the observation of an electron by means of a microscope. One has to bombard an electron with photons to be able to find it. These photons must have a wavelength which is of the same size as the electron itself, or shorter than that, if we are to have exact knowledge of its position. However, wavelength is inversely proportional to momentum, so photons with a short wavelength are also photons with a larger momentum. Some of this will be transmitted to the electron, and it will react by changing its momentum. On the other hand, if one tries not to influence the momentum of the electron by using photons with a very small momentum, then their wavelength will have increased so dramatically that they cannot yield exact information about the position of the electron. This kind of argument was used by Heisenberg to assert that whenever we have exact knowledge of where the electron is we cannot hope to acquire exact knowledge of its velocity, and *vice versa*. He was able to set a limit to how imprecise our knowledge would be in such cases from calculating an equality, called Heisenberg's uncertainty or indeterminacy relation.

At the very same time as Heisenberg completed his paper on the uncertainty principle at the end of February 1927, Bohr seems to have reached what he regarded as a final, coherent understanding of quantum phenomena: for the first time he acknowledged that the quantum postulate, the basis of quantum mechanics, led to the renunciation of a causal space-time description of the atomic phenomena. Heisenberg remembered Bohr's conceiving the foundation of complementarity during a skiing holiday in Norway from the end of February to the middle of March, although it was not until the summer of that year that he seems to have selected the word "complementarity" to refer to this conception.[23]

Chapter VI

1. COMPLEMENTARITY

In September 1927 at the International Congress of Physics in Como commemorating the centenary of Alessandro Volta's death Bohr was ready for the first time in public to disclose his new interpretation of quantum phenomena. In this section we shall lay bare the underlying arguments of this interpretation as they appear in the Como paper and in other of Bohr's earliest essays in order to trace their connections with his philosophical background.

While Bohr was working on the interpretation of quantum mechanics, he was, as we shall see, very much influenced by Høffding's ideas. Over the preceding twenty years Bohr had identified himself so much with Høffding's approach to philosophical problems, his way of describing them and the solutions he put forward that this philosophy had become fundamental to Bohr's own philosophical outlook. We shall discover both Høffding's criterion of reality and his analysis of the subject-object problem in Bohr's development of his conception of complementarity. As noted earlier, Høffding thought that the criterion of reality consists in the permanent connections between our observations, connections that depend on the existence of continuity between the things we observe. And only if we can establish with certainty such connections in our experience do we possess objective knowledge. In other words, the criterion for marking out the existence of the objective sphere is the perceived continuity in our experience which signals the persistence through time of the objects of our perception. But because incompatibilities in the description of our experience occur whenever we face genuine discontinuities, complete unity and connectedness are not always attainable in every area, which means that there will invariably be irrational or unassimilable elements beyond the bounds of intelligibility.

In Bohr's opinion the knowledge of nature that had been obtained through classical physics had been acquired by physicists who, by means of assumptions and experiments, have ordered and synthesized our experience. The aim of classical physics was not to uncover the real essence of phenomena, but to develop methods for ordering and synthesizing human experience in an objective way.[1] Furthermore, the very use of the concept of continuity is characteristic of classical physics. It is possible in the macroscopic world to describe causal connections in space and time, and as a result possible to make a clear-cut distinction between the experimental set-up and the material system, between the physical system used as an instrument of observation and the observed object.

Already here we meet again, now forming part of Bohr's view, one of the essential features of Høffding's theory of knowledge to which we shall later return. Recall that "truth" for Høffding does not consist of certain ideas corresponding with certain facts but of as many ideas as possible connected in a lawful way. We determine truth ourselves by judging whether or not our observations can be coherently ordered and synthesized by our theories. There is no inherent order or essence to be explained. Thus the truth of classical mechanics or of any other theory consists, according to Høffding, in its usefulness in synthesizing and ordering our experiences. And he regarded the possibility of causal description as the criterion by which we determine whether or not our experience is objective.

However, according to Bohr, the objectivity of the descriptions yielded by classical mechanics, of which visualization and causality are the essential features, first underwent a transformation when Planck discovered the quantum of action, and then again through the development of quantum mechanics. The first section following the Introduction of the Como paper, consisting of six paragraphs, is entitled "Quantum Postulate and Causality". These paragraphs illustrate the train of thought which led Bohr to complementarity. The first of them opens with an explanation of why the conditions for description in quantum mechanics differ from those of classical mechanics. Bohr says:

> The quantum theory is characterized by the acknowledgment of a fundamental limitation in the classical physical ideas when applied to atomic phenomena. The situation thus created is of a peculiar nature, since our interpretation of the experimental material rests essentially upon the classical concepts. Notwithstanding the difficulties which, hence, are involved in the formulation of the quantum theory, it seems, as we shall see, that its essence may be expressed in the so-called quantum postulate, which attributes to any process an essential discontinuity, or rather individuality, completely foreign to classical theories and symbolized by Planck's quantum of actions.[2]

If we look at the way Bohr argues we will see that the conclusion is stated before the premisses are given, something which is typical of him. In this and similar passages Bohr seems moreover to equate "atomic phenomena" with "atomic objects", but in other contexts he expresses himself as if he intended to say that the former is the object which appears in the course of observational interaction while the latter is that which in interacting with the measuring instrument gives rise to the phenomena. I shall return to this issue at a later point and for the moment will ignore the ambiguity.

The first premiss in his reasoning is therefore that the classical concepts are indispensable for the description of the experiments which scientists conduct in their search for empirical knowledge of the atom. The exact meaning of that statement is something like this: the apparatus for measurement constitutes a macroscopic object whose construction and function are based on laws successfully accounted for by the classical framework. So the data supplied by an instrument must be understood via a knowledge of its functions, which are described entirely in classical terms. Hence, the empirical data too can only be described by classically defined concepts, and since it is through our knowledge of this data that we can ascribe attributes to the atomic object, we are incapable of forming any description of atomic objects not formulated in terms of classical concepts.

But it is not only because the experimental instruments are constructed and function on the basis of classical physics that "the interpretation of the experimental material rests essentially upon classical concepts". For as Bohr says elsewhere, "It lies in the nature of physical observation that all experience must ultimately be expressed in terms of classical concepts".[3] In fact, if Bohr meant only that experimental data have to be described in terms of classical concepts because the apparatus is such as to conform to classical laws, much modern equipment might not meet this requirement. Indeed Bohr's claim goes deeper than that. What he argues is that it is the very nature of physical observation itself which demands the use of classical concepts.

But what, more precisely, might Bohr have meant by this? The exact nature of any observation seems to be that of bringing order and structure to our experience by constantly imposing the concept of causality on what we perceive in space and time. This is the only warrant we have for claiming that what we experience has objective existence, that is, that the subjective order of experience is isomorphic with regard to an objective order of experience. As late as 1958 Bohr said:

> The description of ordinary experience presupposes the unrestricted divisibility of the course of the phenomena in space and time and the linking of all steps in an unbroken chain in terms of cause and effect.[4]

This is certainly a very Høffdingian thought. The distinction between the subject of perception and the object of perception rests entirely on the ability of the former to organize the content of what is experienced in a way that preserves continuity. This makes possible a notion of the content as an object. Thus, the application of the concept of causality and the concept of space and time to perceptual experience is a precondition for any empirical knowledge of an objective physical world. That is, if it is impossible to establish continuous connectedness in the awareness of sense experience, the distinction between the subjective features of this experience and its objective features become blurred.

In physics that claim is taken a step further, for in the process of physical observation the possibility of describing the state of a physical system in isolation from the state of the instrument of observation by means of classical concepts, and thereby of distinguishing the causal behavior of the observed system from the causal influence of the instrument itself, is grounded on the assumption that the interaction between the object and the instrument can either be disregarded or taken into account. Although Bohr does not make explicit which classical concepts he has in mind when referring to them, he must be thinking particularly of the basic concepts of the classical framework, which represent physical phenomena as causally connected in space and time. Such concepts would include "particle", "wave", "energy", "momentum" and "space-time". So the description of ordinary physical experience presupposes the use of classical concepts which are nothing but refinements of concepts through which classical physics described causal connections in space and time. But we are no longer able to grasp the meaning of the formalism of quantum mechanics or describe the experiments formed within the domain of quantum mechanics unless we use the concepts of classical physics.

The second premiss in Bohr's line of reasoning is that Planck's discovery of the quantum of action was the uncovering of a universal and elementary fact of nature which we have to accept, just as scientists have accepted the empirical fact that light has the same speed in relation to any inertial system. The indivisibility of the quantum of action is a brute fact which cannot be explained away but which has to be accepted as it stands. Furthermore, it provides the atomic processes with an element of discontinuity or individuality (or as Bohr sometimes says, "wholeness", "unity" or "atomicity") which is completely foreign to classical physics. Bohr also argues that although this feature of the individuality or discontinuity of atomic processes is foreign to the classical description, its character of indivisibility is established only in so far as the classical concepts are applied in the description of atomic phenomena.

From the premisses of the indispensability of classical concepts and of the indivisibility of the quantum of action Bohr reaches the conclusion that the use of classical concepts has its limitations. The limits circumscribed by the quantum of action restricts the application of the classically defined concepts within quantum mechanics. In 1929 Bohr summarized his view as follows:

> According to the views of the author, it would be a misconception to believe that the difficulties of the atomic theory may be evaded by eventually replacing the concepts of classical physics by new conceptual forms. Indeed, as already emphasized, the recognition of the limitation of our forms of perception by no means implies that we can dispense with our customary ideas or their direct verbal expressions when reducing our sense impressions to order. No more is it likely that the fundamental concepts of the classical theories will ever become superfluous for the description of physical experience. The recognition of the indivisibility of the quantum of action, and the determination of its magnitude, not only depend on an analysis of measurements based on classical concepts, but it continues to be the application of these concepts alone that makes it possible to relate the symbolism of the quantum theory to the data of experience. At the same time, however, we must bear in mind that the possibility of an *unambiguous* use of these fundamental concepts solely depends upon the self-consistency of the classical theories from which they are derived and that, therefore, the limits imposed upon the application of these concepts are naturally determined by the extent to which we may, in our account of the phenomena, disregard the element which is foreign to classical theories and symbolized by the quantum of action.[5]

Note that Bohr begins by stressing the fact that the limitations of what he calls the forms of perception do not imply that we can do without the ordinary ideas, whatever they are, when bringing order and structure to our sense impressions.

Around 1928–1929 Bohr often spoke about the shortcomings or "the failure of the forms of perception adapted to our ordinary sense impressions".[6] In Danish Bohr always uses the words "*anskuelighed*" and "*anskuelsesformer*", corresponding to the German words "*Anschaulichkeit*" and "*Formen der Anschauung*", of which the latter is used by Kant to denote space and time. "*Anskuelighed*" and "*Anschaulichkeit*" are fairly translated into "visualizability", but in English the standard translation of Kant's "*Formen der Anschauung*" is "forms of intuition". Among the "forms of perception" Bohr includes not only space and time but also causality. He says, for instance, "causality may be considered as a mode of perception by which we reduce our sense impression to order".[7] This use is opposed to Kant's view according to which causality is a category of understanding and not a form of intuition, which space and time exclusively are. However, as we recall, Høffding rejected the sharp distinction which Kant upheld between forms of perception and categories of understanding, regarding both space, time and causality as forms of perception as well as forms of thought or categories of cognition. For Høffding "sensations", "ideas" and "concepts" are all forms in which the content of experience, upon which our cognitive faculties are exercised, appears and is ordered by the mind.[8] Thus Bohr's use of the phrase "forms of perception" is derived from Høffding's expression "forms of thought".[9]

In fact, Bohr's view is quite similar to Høffding's. From Høffding he had acquired the Kantian idea that the content of experience is given through the senses while the form of experience is provided by the act of cognition itself. All sense impressions are conceptually structured in the mind by its representation of the physical phenomena as being in space and time and as being causally connected. If this is the case, then it is the concepts of causality and of space and time which form the "ordinary" ideas with which we cannot dispense. Bohr then goes on to speak about classical concepts instead of ordinary ideas, emphasizing that it is not merely the fact that all instruments of measurement work on the basis of laws established by classical physics that serves to prohibit any abandonment of classical concepts, or makes an alternative impossible, but also the fact that any interpretation of the quantum mechanical formalism has to be expressed in terms of classical physics for it to retain its rootedness in ordinary physical experience. This reasoning appeals to the foundations of the so-called principle of correspondence, and Bohr used it as a philosophical justification of his assertion that classical concepts are necessary for the description of atomic phenomena as well as for any account whatsoever of our physical experience. After he had pointed out that the indivisibility of the quantum of action implies, on the one hand, significant restrictions on the classical theories, Bohr expressed his view in the following way:

> On the other hand, the necessity of making an extensive use, nevertheless, of the classical concepts, upon which depends ultimately the interpretation of all experience, gave rise to the formulation of the so-called correspondence principle, which expresses our endeavors to utilize all the classical concepts by giving them a suitable quantum-theoretical re-interpretation.[10]

Accordingly, he concluded that the problems of an unambiguous description within quantum mechanics could not be solved by substituting some of the classical concepts with a quantum mechanical replacement but, if necessary, by giving them a proper quantum theoretical reinterpretation.

Bohr really never explained why the classical concepts are necessary or indispensable for the description of physical experience. Why is the search for another set of concepts which may describe quantum phenomena better than classical concepts do doomed to be a failure? Admittedly, classical concepts such as "energy", "momentum" and "space-time" have proved their value in the description of every physical phenomenon of the macroscopic world. But this fact does not seem to prove that the experience of atomic phenomena must be described in terms of the same concepts. Indeed, if in our attempt to understand atomic phenomena it turns out that the classical ideal of representation is no longer attainable, we may argue that since the classical concepts were gradually developed to describe macroscopic phenomena they are not well suited to deal with microscopic phenomena. So what is needed now is a new framework which replaces the old concepts altogether, introducing concepts which preserve determinism and eliminate wave-particle dualism. Bohr, however, rejected such an alternative:

> the view has been expressed from various sides that some future more radical departure in our mode of description from concepts adapted to our daily experience would perhaps make it possible to preserve the ideal of causality also in the field of atomic physics. Such an opinion would, however, seem to be due to a misapprehension of the situation. For the requirement of communicability of the circumstances and results of experiments implies that we can speak of well defined experiences only within the framework of ordinary concepts. In particular it should not be forgotten that the concept of causality underlies the very interpretation of each result of experiment, and that even in the coordination of experience one can never, in the nature of things, have to do with well-defined breaks in the causal chain. The renunciation of the ideal of causality in atomic physics ...[11]

Indeed, this passage is interesting for at least two reasons:

First, Bohr makes here a distinction between the *ideal of causality* and the *concept of causality* similar to Høffding's distinction between the principle of causality and the concept of causality. The *ideal of causality*, on the one hand, consists of the idea that a causal representation of physical phenomena in space and time is always a feasible task. It is assumed that every phenomenon is caused by some other phenomenon so that any physical object can be described in terms of its causal connections with other physical objects. This ideal is one which cannot be realized in quantum mechanics. Høffding had already called this ideal into question:

> The principle of causation is a *hypothesis* which has been confirmed only to a certain degree since substantiation of the claim that every phenomenon is brought into an inevitable, let alone, a continuous connection is very remote. It might be said that the principle of causation will never be conclusively confirmed by experience. The principle of causation is an ideal which cannot entirely be attained by cognition.[12]

On the other hand, according to both, the *concept of causality* is necessary for the order and structure of our experience in the sense that only by imposing it on the phenomena we perceive in space and time are we able to obtain knowledge of objects independently of subjective experience. As Bohr says, "In the coordination of experience one can never, in the nature of things, have to do with well-defined breaks in the causal chain". Bohr also emphasizes the fact that the concept of causality underlies the understanding of every experimental result as being an effect of an interaction between the instrument and the atomic system.

Second, in the passage quoted above, Bohr does not mention the indispensability of the classical concepts but of the ordinary concepts or the concept of causality. The reason is obviously that Bohr regarded classical concepts as theoretical refinements of ordinary concepts, i.e., the concept of space and time and the concept of causality. This interpretation is confirmed by the above quotation, where Bohr shifts from talking about ordinary ideas to talking about classical concepts. Thus, the forms of perception are the preconditions of the possibility of sensory experience as well of the meaning of the ordinary language which we use to communicate this experience, including the theoretical refinements of the ordinary words employed in physics. Hence, the argument lying behind Bohr's claim of the indispensability of classical concepts seems to be the following: ordinary concepts are necessary for cognition because they are the concepts or the forms of perception through which the cognizing mind is able to apprehend the outside world. Since the classical concepts have proved during the history of physics to be an elaboration of these forms of perception, the classical concepts are indispensable, too. That is, according to Bohr, the classical concepts refer to the most intrinsic features of physical experience and make explicit the way it has to be structured for the phenomena to be manifested in a cognizable and communicable form. This is the prior assumption underlying Bohr's claim that the classical concepts are indispensable for the description of any physical experience, including "the indivisibility of the quantum of action".

Even though Bohr considered the forms of perception as fundamental for the acquisition of objective knowledge, he also regarded them as abstractions or idealizations. Their abstract nature is brought home, he says, when one looks at the theory of relativity and quantum mechanics:

> In both cases we are concerned with the recognition of physical laws which lie outside the domain of our ordinary experience and which present difficulties to our accustomed forms of perception. We learn that these forms of perception are *idealizations*, the suitability of which for reducing our ordinary sense impressions to order depends upon the practically infinite velocity of light and upon the smallness of the quantum of action.[13]

That is, the invariant distinction between space and time is due to the slowness of movement of ordinary objects of observation just as our usual descriptions of causal connections in space and time depend on the relatively large size of ordinary object of perception. Consequently classical physics, which may be characterized as merely a refinement of the forms of perception, is an idealization or an abstraction. Recall, furthermore, that Høffding maintained a similar point of view. With respect to space, for instance, he claimed that

> *Absolute* or mathematical *space*, the parts of which are completely *homogeneous* and *continuous*, and which has *no* space outside itself, is a mathematical abstraction to which no psychological perception corresponds.... Mathematical space is an abstraction or idealization of experience which is devised for scientific purposes. Our actual experience of space constitutes only an approximation to the properties we attribute to mathematical space, and thus the statements which can be deduced from the character of the mathematical space can only be confirmed approximately by the experience.[14]

Similarly, with respect to continuity Høffding held that it "is an ideal which can only be realized approximately".[15] Thus, Bohr had learned from Høffding that the content of perception may not altogether comply with the forms of perception, which consist of the categories the mind applies to phenomena in order to make them intelligible. Bohr was well-prepared for a situation where the classical framework might fail to apply to our experience of the physical world: in cases where the idealized nature of the classical concepts rendered them inadequate. But, in spite of this fact, Bohr had also been taught that the forms of perception remain the only concepts we have with which we can apprehend that very same experience. So how the incompatibility between the discovery of the quantum of action and the indispensability of the forms of perception should be resolved was the formidable challenge Bohr tried to face by drawing attention to the restricted use of the classical concepts.

In the section of the Como paper following that quoted above, Bohr explains more precisely the nature of the restriction on the application of classical concepts forced upon physicists by the discovery of the quantum postulate.

> This postulate implies a renunciation as regards the causal space-time co-ordination of atomic processes. Indeed, our usual description of physical phenomena is based entirely on the idea that the phenomena concerned may be observed without disturbing them appreciably. This appears, for example, clearly in the theory of relativity, which has been so fruitful for the elucidation of the classical theories. As emphasized by Einstein, every observation or measurement ultimately rests on the coincidence of two independent events at the same space-time point. Just these coincidences will not be affected by any differences which the space-time co-ordination of different observers otherwise may exhibit. Now the quantum postulate implies that any observation of atomic phenomena will involve an interaction with the agency of observation not to be neglected. Accordingly, an independent reality in the ordinary physical sense can neither be ascribed to the phenomena nor the agencies of observation.

At this point Bohr inserted a further passage in the version of the Como paper which appeared in *Nature* where he makes some general philosophical remarks about observation:

> After all, the concept of observation is in so far arbitrary as it depends upon which objects are included in the system to be observed. Ultimately every observation can of course be reduced to our sense perceptions. The circumstance, however, that in interpreting observations use has always to be made of theoretical notions, entails that for every particular case it is a question of convenience at what point the concept of observation involving the quantum postulate with its inherent "irrationality" is brought in.

So, according to Bohr, in the classical framework it is possible to ignore or to account for a possible influence of the measuring instrument on the object under observation. Even the theory of relativity, he says, does not imply a change in this descriptive situation. For the object of observation may be characterized by a set of spatio-temporal coordinates denoting a so-called event and the instrument of observation likewise may be attributed a set of spatio-temporal coordinates designating another event. Both events will coincide independently of the selection of a reference system as it is only causally unconnectable events which do or do not coincide according to the choice of reference. Thus Bohr's argument seems to be that since it is possible in the theory of relativity to ascribe a set of coordinates to both the observed object and the instrument of observation so that these two sets together form an invariant relation, the theory entails the possibility of defining the states of the observed system independently of the states of the observing system.

In opposition to this, it is an essential part of the quantum theoretical framework that interaction between the measuring instrument and the atomic object cannot be ignored due to discontinuity of the quantum of action. Up to the middle of the 1930s Bohr spoke of the finite interaction between object and the instrument of measurement, an interaction determined by the quantum of action in such a way that the atomic phenomena are influenced by us when we observe them. In one place in the paper "The Quantum of Action and the Description of Nature" from 1929 he says about the observation of the atomic phenomena, "We cannot neglect the interaction between the object and the instrument of observation",[16] and elsewhere in the same paper he makes the following statement: "The unavoidable influence on atomic phenomena caused by observing them ...".[17] That this remark does not reflect an imprudence is seen by a similar statement from another paper, "The Atomic Theory and the Fundamental Principles Underlying the Description of Nature" which was also published in 1929: "As we have seen, any observation (in quantum mechanics) necessitates an interference with the course of the phenomena ...".[18] It is therefore beyond doubt, I think, that at that time Bohr's view with respect to observation in quantum mechanics was that the act of observation directly disturbed the atomic object but that this disturbance is uncontrollable due to the quantum of action.

Indeed, talk of disturbance is essentially a *classical* way of speaking which *assumes* all spatially distinct systems exist in distinct mechanical states. If one speaks in this way, then the difference between the classical and quantum mechanical treatment of observation lies in the question of whether the disturbance is "controllable" or not, i.e., whether the perturbation in the object system's state caused by observation can be determined or not. Heisenberg's interpretation of his uncertainty relation by means of the gamma-ray microscope experiment is true in classical terms. The point is then that the perturbation caused by the probing photon will change the electron's momentum in an "uncontrollable" or indeterminate way. Thus we cannot correct for the change of state induced by the observation so as to determine what it would have been if it had remained isolated.

Talk about the disturbance of the atomic object is nevertheless confusing, and Bohr might have inadvertently led his audience into thinking that the limitation of the classical concepts should be understood as a limitation on our knowledge of the classical states of a quantum mechanical system. For if it is merely the case that we disturb the atomic object when observing it, it would appear that the object might be in a definite state before as well as after the observation was made, although we in principle are cut off from knowing anything empirically about these states because of Planck's constant. This was, probably, the way many of Bohr's critics saw the situation and the way they interpreted his talk of disturbance. The charge that this was what Bohr seemed to mean might have been justified further if Bohr at the same time had been strongly biased in favor of a representational theory of knowledge. However Bohr, like Høffding, was keenly opposed to all the ingredients of that epistemology. Explicitly or implicitly he rejected an *ontological* defence of an external world, a correspondence theory of truth, a picture theory of knowledge, strong objectivism, and a sharp distinction between subject and object. Instead he adhered to an *epistemological* defence of the external world, a coherence theory of truth, a non-picturing theory of knowledge, weak objectivism, and to there being a blurred distinction between subject and object. Since Bohr's epistemology differed from that of most physicists at the time he did not look with any suspicion on phrases like "the disturbance of atomic phenomena". When Einstein, Podolsky and Rosen in 1935 introduced an argument in favor of the existence of definite, unobservable quantum states, it probably took him by surprise as it forced him to change parts of his terminology and arguments.

Now, because of the individuality of the quantum of action and the resulting uncontrollable interaction between the measuring instrument and the atomic object, the problem of not being able to define a classical state of an atomic system is solvable only if we limit the application of a causal description of the system in space and time. As Bohr says, "Only by a conscious resignation of our usual demands for visualization and causality was it possible to make Planck's discovery fruitful in explaining the properties of the elements on the basis of our knowledge of the building stones of atoms".[19] Similarly, writing about the success of the description of the atomic structure as a natural generalization of classical mechanics: "This goal has not been attained, still, without a renunciation of the causal space-time mode of description that characterizes the classical physical theories which have experienced such a profound clarification through the theory of relativity".[20] Consequently, we cannot acquire the same kind of objective knowledge that was the ideal in classical mechanics, because the quantum mechanical system cannot be made to conform to the concept of causal relations together with a space-time mode of description. An atomic system can no longer be assigned both a certain energy and momentum and, at the same time, a certain spatio-temporal locus because of the quantum of action. We can say therefore that an *incompatibility* made its appearance in our customary description of physical experience when the quantum of action was discovered.

Quite in agreement with Høffding's characterization of an element that cannot be made to conform to a causal description as an *irrational* element not subsumable by the categories of the mind, we find that Bohr on several occasions terms the incompatibility in question an *irrational* element in relation to the classical causal description. Bohr uses, for instance, this designation in the above quoted passage which was inserted in the *Nature* version of the Como paper, another passage which seems to reflect his discussions with Høffding during 1927–1928. In 1929 he again talked about the quantum of action as an irrationality: "In the quantum theory we meet this difficulty at once in the question of the inevitability of the feature of irrationality characterizing the quantum postulate".[21] And, just to mention one further instance of its occurrence, in the Introductory Survey to the collection of papers *Atomic Theory and the Description of Nature* of 1929, Bohr states the same idea in referring to the Como paper: "It is maintained in the article that the fundamental postulate of the indivisibility of the quantum of action is itself, from the classical point of view, an irrational element which inevitably requires us to forgo a causal mode of description".[22] So with the quantum of action we face what is a restriction on Høffding's criterion of reality, with the result that our knowledge of atoms remains "incomplete" owing to the presence of the irrational element.

To repeat, the finite magnitude of the quantum of action constitutes the reason why the quantum mechanical system interacts with the experimental instrument in such a way that one cannot fully determine or control this interaction, and owing to which it is no longer possible to describe a causal connection between the system under observation and instrument used for the observation. Bohr therefore concludes that "independent reality in the ordinary physical sense can neither be ascribed to the phenomena nor the agencies of observation". In other words, without a possible causal description of the interaction between the atomic phenomenon and the measuring instrument we are robbed of an objective and non-arbitrary criterion serving to separate the system from the means of observation and to justify the meaningfulness of the ascription of "an independent reality" to the observed phenomenon. In 1929 Bohr amplified what he meant by "independent reality" in the jubilee paper "The Quantum of Action and the Description of Nature":

> The very recognition of the limited divisibility of physical processes, symbolized by the quantum of action, has justified the old doubt as to the range of our ordinary forms of perception when applied to atomic phenomena. Since, in the observation of these phenomena, we cannot neglect the interaction between the object and the instrument of observation, the question of the possibilities of observation again comes to the foreground. Thus, we meet here, in a new light, the problem of the objectivity of phenomena which has always attracted so much attention in philosophical discussion.[23]

The "old doubt" Bohr here speaks of is indeed the skepticism with regard to the universal application of the forms of perception which formed an essential element of Høffding's epistemology. If there are situations like those in atomic physics where the ideal of causality cannot be met, we are forced to dispense with an unambiguous criterion of reality which can be applied to subjective experience, and for that reason the objectivity of the experienced phenomena is called into question.

Later in the same paper Bohr returns to the problem of objectivity, referring directly to the final statement of the Como-paper which, I have argued, was probably added to the *Nature* version after discussions with Høffding. Here he writes:

> At the conclusion of the paper referred to, it was pointed out that a close connection exists between the failure of our forms of perceptions, which is founded on the impossibility of a strict separation of phenomena and means of observation, and the general limits of man's capacity to create concepts, which have their roots in our differentiation between subject and object.[24]

A relatively sharp distinction between the observed object and the observing subject is a presupposition of our ordinary concept of observation, and this distinction is a result of the ability of cognition to subsume the subjective order of experience under the forms of perception. Given the impossibility of this occurring, no objective order of experience can be established and the mind cannot form a concept of the content of experience as an object. This applies too in the case of our observation in the realm of the quantum of action because "every observation can ultimately be reduced to our sense perception". So what Bohr means by the difficulty of drawing a distinction between atomic phenomena and their observation is that the breakdown of the forms of perception in quantum mechanics, owing to the inseparability of the interaction between the objects of investigation and the instruments for observation, is a special case of the general epistemic situation in which the distinction between subject and object becomes blurred. This happens in situations in which continuity between "our sense perceptions" fails to obtain. So in cases where continuity provided by the forms of perception has no or only restricted application to sense experience, the concept of an independent, objective world breaks down.

In the paper "The Atomic Theory and the Fundamental Principles underlying the Description of Nature", written in the same year as the jubilee paper, Bohr again expresses the idea that the restriction of causal descriptions in space and time of atomic phenomena gives rise to a questioning of the objective existence of atomic phenomena independently of observation:

> The discovery of the quantum of action shows us, in fact, not only the natural limitation of classical physics, but, by throwing new light upon the old philosophical problem of objective existence of phenomena independently of our observation, confronts us with a situation hitherto unknown in natural science. As we have seen, any observation necessitates an interference with the course of the phenomena, which is of such a nature that it deprives us of the foundation underlying the causal mode of description. The limit, which nature herself has thus imposed upon us, of the possibility of speaking about phenomena as existing objectively finds its expression, as far as we can judge, just in the formulation of quantum mechanics.[25]

The finite size of the quantum of action is to blame for the fact that the atomic system interacts with the measuring instrument in such a way as to make the extent of the interaction indeterminable. Since a causal mode of description of atomic interaction, such as that demanded by the criterion of reality, can no longer be sustained, this gives rise to the notorious conflict between the inseparability of object and instrument owing to the quantum of action and the separability of object and means of measuring required by the concept of observation.[26]

But how can the incompatibility brought into being by the claim of the quantum of action and that of the separability of the phenomena under investigation and the measuring instruments be solved? Bohr's answer is that the separation of the object of observation and the apparatus that renders it observable is partly an arbitrary one. If observation is to be possible a sharp distinction is required between the subject making the observation and the object being observed. There must be an apprehending subject who is not at the same time the object apprehended. Furthermore, since every observation is reducible to sense perceptions, our experience has to be interpreted by theoretical notions in the form of classical concepts if the content of sense perceptions is to be regarded as a result of an observation. Like Høffding before him, Bohr rejected, in opposition to the logical positivists, the idea of neutral sense data: all experience is highly theory-laden by being structured and ordered by a conceptual framework, and as long as continuous connections between our sense perceptions persist they guarantee that our observations are of an independent object. To generalize one might say that the forms of perception, according to Bohr, are necessary because they embody the distinction between subject and object.

With respect to quantum mechanics this analysis of observation entails, on the one hand, that a sharp distinction between a phenomenon and the system of observation is a precondition for talking about observing an object independently of the means of observation. The instrument used for measurement cannot be part of the object of investigation for an observation to be made. On the other hand, if the instrument is to be treated as an instrument and not as an object, then the measurement interaction with the object is indeterminable from the point of view of observation, because the interaction can only be determined if the measuring device is considered simultaneously as an instrument and an object, which is logically impossible. But since the disturbance prevents us from extending causal descriptions to isolated states of the atomic system we must, pragmatically, make a more or less arbitrary distinction between an *observed* object and the instrument of observation by treating the observed object as being conceptually distinct from the instrument, and how this is done will depend upon the aims and interest of the scientists in a given situation. Thus sense perceptions which form part of the process of observation are not those of an object having independent reality but of an observed object.

The separability of the observed object and the instrument is not entirely arbitrary. For as Bohr says, "The concept of observation is in so far arbitrary as it depends upon which objects are included in the system to be observed". What is arbitrary, then, is that the demarcation may be made either between the observed object, consisting only of the microscopical object, and the instrument, or between an observed object consisting of the microscopical object along with an instrument, or a part of one, and a further instrument. Where to place the demarcation is "a question of convenience", but one has to make a choice. The choice of a distinction is necessary in order to regard the act of the measuring instrument as an act of the observation of an object. And, as we have seen, in order to describe the outcome of an observation precisely it has to be done in classical terms, because "only with the help of classical ideas is it possible to ascribe an unambiguous meaning to the results of observation".[27] It is the classical concepts which refer unambiguously to the properties of an object as observed. These concepts are indispensable for describing both the functions and results of the measuring instrument because the object-instrument distinction is also an inherent feature of the use of classical concepts. The use of classical modes of description to account for both the construction and manipulation of the measuring instrument and the result of the experiment thereby allows us to separate the instrument from the observed object, in spite of their causal inseparability.[28] The object itself, whether or not it is a composite object, including the interaction of the measuring instrument, lies outside the range of classical physics, since it can only be described by a quantum mechanical framework with the restricted application of classical concepts, for which reason no result of the experiment "can be interpreted as giving information about independent properties of the objects".[29]

Bohr thus came to the conclusion that the separability in quantum mechanics is in some sense fundamental since it is what makes classical concepts necessary:

> The essentially new feature in the analysis of quantum phenomena is, however, the introduction of a *fundamental distinction between the measuring apparatus and the objects under investigation.* This is a direct consequence of the necessity of accounting for the functions of the measuring instruments in purely classical terms, excluding in principle any regard to the quantum of action.[30]

However, although the distinction is fundamental in an epistemological sense it is still merely pragmatically justifiable because we cannot theoretically define the state of the system under investigation independently of the state of the measuring instrument as we can do within the classical framework. Furthermore, one has to remember that in principle any physical system can be described quantum mechanically. This explains, perhaps, Bohr's obscure remark in the Como paper, quoted above, claiming that a consequence of the interaction between the atomic phenomena and the measuring instrument is that "an independent reality can neither be ascribed to the phenomena nor the agencies of observation". But then we are debarred from treating the instrument as an instrument. What we can do to surmount the difficulty is to appeal to the size of the instrument. An instrument of observation must be a macroscopic object whose interaction with another object, according to the classical framework, is so vast relative to that measured by Planck's constant that it is possible both to give a classical account of its relative position and momentum as well as the interaction with another macroscopic object.[31] And only because we are thereby justified in applying a classical description to the functions and results of the instrument is it possible for us to claim that we are making an observation of an object, an assumption we want to make in order to ensure that the results of the experiment are epistemically useful.

Thus, Bohr's considered opinion of the problem of observation in atomic physics seems to be as follows: The language of classical physics is the only language we have in which to talk about the physical world as an objective and independent reality in an unambiguous and rational way. The irrational element that has arisen in physical experience upon the discovery of the quantum of action, appears precisely as such against the background of the classical concept of the world. Therefore we can neither obviate the difficulties nor dispense with classical concepts: they are necessary because they embody the assumption of object-instrument separability by forming the basis for the demonstration of the presence of those causal relations in space and time that constitute the criterion for reality. When all is said and done, it is the use of the space-time concepts and the conservation theorems for energy and momentum that characterize the classical description of continuity – which is necessary for us to be able to synthesize and describe our physical experience, which is also to say, the results yielded by the measuring instruments, in an unambiguous and objective manner.[32] Nevertheless, the discovery of the quantum of action imposes a limitation on the notion of an objective distinction between an object and an instrument due to the fact that the interaction between them is uncontrollable and therefore unaccountable in terms of an unrestricted use of causal and spatio-temporal modes of description. The dilemma which arises owing to the causal inseparability of object and instrument due to the quantum of action, on the one hand, and owing to the separability of object and instrument demanded by the concept of observation, on the other, is solvable only if we discriminate between the notions of "independent object" and "observed object", depending on whether we are dealing with a quantum mechanical or classical context of description.

In the following passage of the Como paper Bohr explains in greater detail how in his view a partial renunciation of causality and visualizability and the introduction of a partial arbitrariness in the distinction between object and measuring instrument contributes to a consistent description of atomic system:

> This situation has far-reaching consequences. On one hand, the definition of the state of a physical system, as ordinarily understood, claims the elimination of all external disturbances. But in that case, according to the quantum postulate, any observation will be impossible, and, above all, the concepts of space and time lose their immediate sense. On the other hand, if in order to make observation possible we permit certain interactions with suitable agencies of measurement, not belonging to the system, an unambiguous definition of the state of the system is naturally no longer possible, and there can be no question of causality in the ordinary sense of the word. The very nature of the quantum theory thus forces us to regard the space-time co-ordination and the claim of causality, the union of which characterizes the

classical theories, as complementary but exclusive features of the description, symbolizing the idealization of observation and definition respectively.

Here Bohr first refers to the classical situation in which it is possible to define the state of an isolated system on the basis of the observation of a given state. For it is always taken for granted in classical mechanics that the effect of the very process of investigation of the system could be neglected or could be taken into account. But the situation is changed in quantum mechanics. Because of the uncontrollable nature of the connection between atomic phenomena and their observation it is impossible in quantum mechanics to ascribe simultaneously properties which are necessary for an exhaustive description of an object according to classical theory. The precondition of a classical description consists of an observation and a definition. In order to define the state of an atomic system to which we can apply the concept of causality provided by the theorems of energy and momentum conservation, it is required that the system be isolated from its surroundings. But in that case we are unable to determine empirically the space-time coordinates necessary to define the state of the system, for an observation is required for us to be able to make the relevant ascriptions. If, on the other hand, we observe the system in order to determine the space-time coordinates, we are not in a position to give a precise definition of the state of the isolated system that would provide a continuous description of the future states of the system. The consequence is, Bohr says, that causal and space-time modes of descriptions are complementary.

There has often been considerable uncertainty as regards the meaning of complementarity, for which Bohr is partly to blame, since he never defines what he exactly means by this term. In the Introduction to *Atomic Theory and the Description of Nature* Bohr says that the quantum of action "forces us to adopt a new mode of description designated as *complementary* in the sense that any given application of classical concepts precludes the simultaneous use of other classical concepts which in a different connection are equally necessary for the elucidation of the phenomena".[33] This is, perhaps, the best and clearest exposition of what he meant by complementarity. It is a mode of description in quantum mechanics according to which we cannot ascribe a well-defined value of energy and momentum to an atomic system at the same time as attributing to it a precise value in space and time. If we require an accurate application of the one set of concepts, then we must invariably forgo the precise application of the other set of concepts. However, only part of what is meant by complementarity is expressed by saying that the simultaneous application of two different sets of concepts are mutually exclusive or incompatible. The other is that the joint use of these two sets is necessary to provide us with a complete description or characterization of the system.

Thus, the complementarity of spatio-temporal and causal concepts reflects the restrictions on our ability to determine the exact values of the states of an atomic system on the basis of measurement. The limitations of measurability or observability are precisely what is expressed by Heisenberg's uncertainty relations, Bohr argues.[34] The position of a particle in space and time can be observed to have an exact value only if its energy and momentum at the same time have a corresponding inexact value, or *vice versa*, and this is due to the indivisibility of the quantum of action. It is the fuzziness of the measurement of either the position in space and time of the atomic system or of the energy and momentum of the system which precludes the simultaneous application of the classical concepts and thereby prevents us from describing the trajectory of the system in causal terms. Through the analysis of various thought experiments Bohr was able to demonstrate that the experimental conditions for measuring the spatio-temporal position of an object were incompatible with those for measuring its momentum and energy. He shows, for instance, that if the position of the object is to be measured with such precision that the uncertainty relation is of consequence, then the position of the measuring instrument must be known with the same high degree of exactness. But this condition is satisfied only if the measuring instrument is fixed rigidly to the experimental set-up which defines the spatial reference frame. On the other hand, the momentum of the object can only be measured if the measuring device is loosely connected to the apparatus defining the spatial reference frame. Measurement of the momentum demands that the momentum of the measuring device is known both before and after its interaction with the object, so that the momentum of the object can be calculated from the change in the momentum of the instrument by employing the conservation law of momentum. However, the exact measurement of the change in the momentum of the instrument requires a loose connection to the entire set-up which defines the spatial reference frame. This condition is incompatible with the condition of fixity necessary for a position measurement. Experimental possibilities for a precise momentum measurement and for a precise position measurement are therefore mutually exclusive.

There is no doubt that Bohr thought that the reason why the concepts of causality and space-time are complementary is because they are empirically or epistemically incompatible. But whether or not he also believed that they are conceptually or ontologically incompatible is something we shall discuss in Chapter VII.

A complementary relation similar to that between the kinematic and dynamic concepts applies to the wave-particle aspect of atomic phenomena; in certain circumstances atomic phenomena display wave-like properties and in other situations they display particle-like properties. Nevertheless, here too the restriction applies that these incompatible descriptions rest on mutually exclusive experimental situations. In the Como paper, shortly after the passage in which he formulates kinematic-dynamic complementarity, Bohr states:

> The two views of the nature of light are rather to be considered as different attempts at an interpretation of experimental evidence in which the limitation of the classical concepts is expressed in complementary ways. ... Just as in the case of light, we have consequently in the question of the nature of matter, so far as we adhere to classical concepts, to face an inevitable dilemma which has to be regarded as the very expression of experimental evidence. In fact, here again we are not dealing with contradictory but with complementary

pictures of the phenomena, which only together offer a natural generalization of the classical mode of description.[35]

Around 1926 wave-particle duality was a crucial issue for Bohr in his groping towards the notion of complementarity, whereas kinematic-dynamic duality was regarded by him merely as a consequence. But in the Como paper Bohr begins his exposition of complementarity by pointing to kinematic-dynamic complementarity as a consequence of the incompatibility of causal and space-time modes of description before turning to wave-particle complementarity. Now Bohr seems to believe that wave-particle complementarity is also a result of "the impossibility of a causal space-time description of the light phenomena" as well as of the impossibility of a continuous description of particle phenomena. The complementary use of the classical pictures or models, as one might prefer to call them, of particles and waves as representing both matter and light are considered a conceptual consequence of the limitations of classical concepts, resulting from the exclusive application of causality and space-time coordination. Thus, on the one hand, propagation of light in space and time is associated with the wave picture with its superposition principle, but in the case of an exact measurement of the propagation in space and time, energy and momentum, owing to the uncertainty relation, are imprecisely measured, and "we are confined to statistical considerations" in attempting a space-time description of light propagation, divested as we are of "the fulfillment of the claim of causality".[36] On the other hand, the energy and momentum of light are correlated with the particle picture and, owing to the uncertainty relation, if they are exactly measured the particle's position in space and time cannot be precisely determined but involves a renunciation of the space-time description of light which belongs to the wave picture.[37]

However, sometimes the wave picture can be, contrary to what Bohr seems to indicate here, correlated with the measurement of momentum and energy while spatio-temporal measurement is connected with the particle picture. If a certain momentum or the energy of an object is measured one may, on the basis of Einstein's and de Broglie's equations, assign a certain wavelength or frequency to the object, and this in turn gives rise to a wave picture in which the object is represented as spread out in a region of space. But if the measurement of a certain spatio-temporal position is carried out the location of the object evokes a particle picture. Although this formal approach to the two kinds of model does not seem to reflect Bohr's point of view, which was based on the interpretation of physical experiments, Murdoch has in his book on Bohr argued, and rightly I think, that neither the wave picture nor the particle picture is a consequence of the limitation on the application of either momentum or position but that wave-particle complementarity and kinematic-dynamic complementarity are logically distinct notions.[38] It is difficult, if not impossible, to see how the duality of the wave-particle pictures can be derived from the separation of causality and space-time coordination, allowing both theoretical and experimental correlations. If this is the case, it has far-reaching implications for potential criticism of one kind of complementarity without having any influence on the other kind, something which both Bohr and many of his opponents have overlooked.[39]

In spite of the fact that Bohr saw the wave-particle duality as an integral part of the framework of complementarity there are, as Murdoch has also observed, some indications that Bohr did not think that the wave picture of matter and the particle picture of radiation had the same realistic significance as the pictures of matter and radiation regarded as particles and waves, respectively.[40] The reason is quite clear: the experimental evidence, to take one example, for matter waves is as reliable as that for considering the propagation of light as waves, and such evidence constitutes the reason why Bohr believed that the classical pictures of particle and wave are equally necessary for the theory of matter and the theory of radiation, but

> we must bear in mind that the application of matter waves is limited to those phenomena, in the description of which it is essential that the quantum of action be taken into account and which, therefore, lie outside the domain where it is possible to carry out a causal description corresponding to our customary forms of perception and where we can ascribe to words like "the nature of matter" and "the nature of light" meanings in the ordinary sense.[41]

What Bohr means is, of course, that the classical pictures of light radiation as a wave disturbance in a field and matter as particles have, after all, taken their rise from experimental situations where the criterion of reality applies, and these models have therefore a different ontological status than that of the models of matter waves and radiation as corpuscles which are drawn from experimental evidence to which the criterion of reality fails to apply.

Nevertheless, Bohr regarded complementarity as a more general framework for the description of reality than the classical framework of causality. Quantum mechanics, he says, "Forces us to replace the ideal of causality by a more general viewpoint usually termed 'complementarity'".[42] That is, deterministic predictions of the behavior of an individual atomic system which presupposes causal relations in space and time are now being replaced by statistical, or rather probabilistic, predictions of the likely results of further measurements on that system. A similar point of view concerning complementarity is expressed in his paper "Discussion with Einstein on Epistemological Problems in Atomic Physics", in which Bohr makes the following remark about his 1927 paper: "The trend of the whole argumentation presented in the Como lecture was to show that the viewpoint of complementarity may be regarded as a rational generalization of the very ideal of causality".[43] Such statements can only be understood in the light of Bohr's conception of continuity as the criterion of reality in relation to the difficulty of clearly distinguishing between the atomic object and measuring instrument in quantum mechanics. For in both cases Bohr mentions complementarity only as an extension of the *ideal* of causality, not the *concept* of causality. This is because it is merely the ideal of causality which we have to abandon in accepting the quantum postulate.[44] Contrary to the ideal of causality, the concept of causality is an essential component of the complementary framework in virtue of the fact that causal and space-time descriptions are mutually exclusive. Causality as a concept is still necessary, together with the concept of space and time, for bringing order and structure to our sense experience and thereby establishing the fact that we are dealing with objective phenomena. But the quantum postulate puts limits to their universal application in the form of restrictions on the use of classical concepts such as space and momentum. What we have is a curtailment of the use of the criterion of reality, and for this reason we cannot use the classical concepts to refer without ambiguity to the state of an observation-independent atomic system. As Bohr himself expresses it, "The notion of complementarity serves to symbolize the fundamental limitation, met with in atomic physics, of the objective existence of phenomena independent of the means of their observation".[45] That is, when applied to microscopic phenomena, the criterion of reality does not justify confidence in the objectivity of unperturbed or isolated states of the system.

There are two central elements in Bohr's interpretation of quantum mechanics. The principle of correspondence ensures the legitimacy of the use of the classical concepts in descriptions of all physical experience in spite of the discontinuity of the quantum of action, whereas the framework of complementarity limits their use in relation to the classical theories in order to accommodate the discontinuity. But this limitation of the use of classical concepts is tantamount to a restriction on the application of the criterion of reality. Hence, complementarity is concerned with the objectivity of the descriptions of the observations of the atomic system rather than with the objectivity of the descriptions of the isolated system itself.

2. QUANTUM MECHANICS AND PSYCHOLOGY

The strands of Bohr's thought that we have traced and found to be in accord with Høffding's epistemology may be briefly summarized as follows: The criterion of objectivity and reality of a thing lies in the possibility of establishing a causal relation – a condition which was fulfilled in classical physics. The finite magnitude of the quantum of action is the reason why we can no longer use the classical ideal of causality, for the causal space-time mode of description of atomic phenomena is precisely what the quantum of action precludes. Here we encounter an irrational element in our cognition of atoms. As we cannot control the interaction between the phenomenon and the experimental instrument – because of the finite magnitude of the quantum of action – it becomes impossible to distinguish between the phenomenon and the means by which it is observed. Thus the question arises as to whether or not the phenomena exist independently of our observations – and here we meet the old philosophical problem of the relation between subject and object which, according to Bohr, is the problem at the core of epistemology.[46]

At this stage in his reflections, Bohr evidently turned to the analysis of the conditions for the observation of psychological phenomena and indicated that this analysis bore similarities to that of the conditions for the description of quantum mechanical phenomena. This analogy between psychology and atomic physics seems to have originated in Høffding's thinking. From Bohr's address in 1932 to the participants at the International Psychology Conference in Copenhagen, quoted in Chapter III, we have learned that Bohr believed that the "individuality" of the measurement interaction parallels the indivisibility of subject and object in the investigations of psychological phenomena, and that this idea is clearly one that had originated with Høffding. I also argued in that chapter that the analogy between quantum mechanics and psychology was something which Høffding treated explicitly in a short essay which he sent to Bohr in the summer of 1928, after having called Bohr's attention, when preparing the Como paper for publication in the winter of 1927–1928, to the general epistemological problem of distinguishing between subject and object in cases where observation unavoidably interacts with what is observed. Høffding and Bohr discussed the analogy during the summer before Bohr began writing his contribution to the Planck Jubilee issue, "The Quantum of Action and the Description of Nature", in which Bohr for the first time referred to that analogy. As these discussions continued throughout the remainder of 1928 and in 1929 (which we know they did from Høffding's letters to Meyerson), there is good historical evidence for believing that this paper and the other paper from 1929 were written under Høffding's direct influence and bore the imprint of his thought.

Thus in the Planck Jubilee paper, published on 28th June 1929, Bohr had at the last moment decided "to leave all physics out and stick to pure philosophy", as he declared in a letter to Pauli.[47] There he pointed out the familiarity of complementary modes of thinking in psychology: "The necessity of taking recourse to a complementary, or reciprocal mode of description is perhaps familiar to us from psychological problems".[48] In the same year Bohr emphasized the analogy between problems of observation in quantum mechanics and psychology in another paper, "The Atomic Theory and the Fundamental Principles Underlying the Description of Nature", which Bohr delivered as a talk at a gathering of Scandinavian scientists between 26th and 31st August. There he said:

> ... the linkage of the atomic phenomena and their observation, elucidated by the quantum theory, does compel us to exercise a caution in the use of our means of expression *similar to* that necessary in psychological problems where we continually come upon the difficulty of demarcating the objective content.[49]

Bohr obviously saw similar situations in both quantum physics and psychology with respect to the difficulties of separating the knowing subject from the object it knows. This is clearly expressed in a parallel passage from the "Introductory Survey", also written in 1929:

> The impossibility of distinguishing in our customary way between physical phenomena and their observation places us, indeed, in a position quite similar to that which is so familiar in psychology where we are continually reminded of *the difficulty of distinguishing between subject and object.*[50]

Thus, atomic physics and psychology exemplify the same epistemological problem, a problem which psychologists had been acquainted with for a long time.

In 1948 Bohr recalled the time when he became aware of the common epistemological ground shared by atomic physics and psychology:

> ... quantum theory presents us with a novel situation in physical science, but attention was called to the very close analogy with the situation as regards analysis and synthesis of experience, which we meet in many other fields of human knowledge and interest. As is well known, many of the difficulties in psychology originate in the different placing of the separation lines between object and subject in the analysis of various aspects of psychical experience.[51]

The person whose attention "was called to the very close analogy" between atomic physics and psychology "as regards analysis and synthesis of experience" was undoubtedly Bohr himself, as he was informed by Høffding in the course of their discussions about the epistemic similarities in the two fields. This suggestion is supported by Bohr's very peculiar choice of terminology. Bohr's use of the terms "analysis" and "synthesis" seems to be highly idiosyncratic until one realizes that he must have borrowed these phrases from Høffding. Høffding very often employed, just as Bohr did, the terms "analysis and synthesis" to describe methods by which human knowledge is increased. "Analysis" signifies the movement in the line of reasoning from the complex whole to the singular elements, while "synthesis" denotes the act of bringing the elements together in a coherent way. Analysis and synthesis are in Høffding's methodology the counterparts of discontinuity and continuity in his epistemology, and so they are in Bohr's philosophy as well. Reflection breaks the immediately given experience into items, whereupon it assembles the separate parts into new wholes by bringing the elements under the concept of continuity. Høffding would say that through analysis we make the content of experience subjective and through synthesis we make it objective. This also explains why Bohr compares the methodological distinction between the analysis and synthesis of experience with the epistemic subject-object distinction in the passage just quoted by saying that the latter distinction was a problem for the analysis of mental experience.

An awareness of the problems of distinguishing between subject and object is indeed an intrinsic feature of Høffding's thought, since, as we have seen, he laid great weight on the incompatibility of a distinction between subject and object in contexts where the criterion of reality fails to apply. According to Høffding, in every act of cognition it is necessary to distinguish between a subjective and an objective element, each of which implies the other, although they manifest themselves in this relation to a different degree. The distinction may be defined by saying that what can be described under the law of continuity is objective and what breaks the law of continuity is subjective. But then dogmatic and speculative philosophers and scientists are inclined both to forget, Høffding argued, that the concept of continuity is in itself a form of thought which the subject tries to establish in what is given as the content of experience, as well as disregarding the fact that not all qualia and spatial and temporal differences are merely a product of the conceptual activity of the subject itself. These differences form the material for the act of cognition, although the form with which and the degree to which they manifest themselves are due to the mind's own involuntary conditions. Thus, according to Høffding, we always set an objectively determined subject in contradistinction to a subjectively determined object. And the irrational element appears because one cannot isolate a "pure" object or a "pure" subject, whose properties can be derived from the concept itself. Instead we have a series of alternating subjects and objects of the following type: $S_1\{O_1\{S_2\{O_2\{S_3 \ldots$, in which the difference between the subject and the object appears anew irrespective of whether we have once given an objective characterization of the features pertaining to the subject and a subjective characterization of the features pertaining to the object.[52]

When Høffding discussed the conditions for the description of psychological phenomena, as pointed out earlier, he time and again emphasized the peculiar circumstance that a state of consciousness may undergo change when under observation because the observing subject indirectly affects the object that it observes, and thus the observed object, *in casu* a state of consciousness, acquires a different nature. This is due to the impossibility of distinguishing between the observing and the observed parts of the mind, which cannot be entirely separated. Likewise, in experimental psychology this means that attention should be paid to the specific rules and conditions under which the experiment is carried out. Regardless of whether it is a case of self-observation or of psychological experiments conducted on others, one affects the state of consciousness that one seeks to describe. Moreover, in situations of this sort it will often be quite impossible for other significant aspects of consciousness to co-exist with that under examination.

Høffding claimed this to be an important feature of the description of mental states in man since we are here confronted with mutually exclusive descriptions – not always so in a logical sense but rather in that they simply cannot co-exist as psychological states. Høffding pointed out that it is impossible to study the psychology of one's own will and take action at the same time. And yet descriptions of both aspects of willed acts are important in a comprehensive picture of one's own conscious mental life. Emphasis on this circumstance was perhaps even more to the fore when Høffding discussed the psycho-physical problem. He wrote that physiological and psychological considerations, although incompatible, complement each other in such a way that mind and body must be considered as two sides of what is in fact a unity. Here we cannot make do with one set of descriptions, a physiological set or a psychological set, for these two sets are not individually exhaustive; jointly they are so.

Given the light shed on it by Høffding's epistemology of psychology, Bohr's thinking in this area can be set forth as follows: In psychology the situation encountered is patently the same as in quantum mechanics, for in both fields there is the difficulty of delimiting the objective content of what is observed inasmuch as the knowing subject influences the object known. Thus in "The

Quantum of Action and the Description of Nature" Bohr wrote:

> The unavoidable influence on atomic phenomena caused by observing them here corresponds to the well-known change of the tinge of the psychological experiences which accompanies any direction of the attention to one of their various elements.[53]

When Høffding read this essay, he made, as was his custom, notes on its content and copied out the statement which we have just cited, putting with it a reference to the tenth edition of his own *Psykologi* from 1925.[54] Surely, no statement better describes than does Bohr's the psychological situation that Høffding had called attention to in a passage of the beginning of that book, a passage which has been quoted above on page 97. To see just how close Bohr's view really is to Høffding's, let us look at a part of the passage once more:

> Attention in itself changes the state to which it is directed. This happens so much more easily as the observing and the observed parts of consciousness cannot in reality be kept entirely apart. The expectation of finding certain thoughts or feelings in ourselves can, without our noticing anything, cause the state to be changed in the expected direction.

The passage can also be found in the 1898 edition of Høffding's *Psykologi*, which Bohr read in preparation for his examination in propaedeutic philosophy, "*Filosofikum*". However, it is very conceivable that these were the descriptive problems of psychology touched upon by Høffding in the missing essay on the principle of causation in the modern electron theory.

The agreement between the thinking of Høffding and that of Bohr stands out even more clearly in the following passage from Bohr's writings, where he describes the nature of the subject-object problem in psychology.

> The epistemological problem under discussion may be characterized briefly as follows: For describing our mental activity, we require, on one hand, an objectively given content to be placed in opposition to a perceiving subject, while, on the other hand, as is already implied in such an assertion, no sharp separation between object and subject can be maintained, since the perceiving subject also belongs to our mental content. From these circumstances follows not only the relative meaning of every concept, or rather, of every word, the meaning depending upon our arbitrary choice of view point, but also that we must, in general, be prepared to accept the fact that a complete elucidation of one and the same object may require diverse points of view which defy a unique description.[55]

For Bohr, as for Høffding, the subject-object problem in atomic physics is similar to that met with in psychology because both fields exemplify the same underlying epistemological argument. But what exactly does this argument look like? It starts out from the fact that in classical physics it has been possible to separate in a fairly clear way that which is objectively known from the knowing subject by applying the criterion of reality to phenomena experienced through sense-impressions. In both atomic physics and psychology the criterion of reality cannot be applied to the same extent to the object under investigation because of the unity of the quantum of action and the unity of consciousness, respectively. One consequence that follows from this is that the observation of an atomic state or a mental state disturbs the observed phenomenon so that one is unable to ascribe such states to any object or mind independently of the perception of it. In both types of enquiry the act of observation itself modifies the states which are under examination, the reason for this being that the state qua observed state constitutes a whole in respect of which we cannot sharply distinguish subject from object. Given the existence of this "whole" one cannot apply the criterion of reality to the subject and object relation itself as the law of continuity is the condition of this distinction, and the independence of the object is thus called into question.

Nevertheless, it may be objected that something has gone wrong here. For if the world had turned out to be classical, would such an "irrationality" as the quantum of action have arisen? Høffding seems to need an affirmative answer. But Bohr with the quantum postulate doesn't. The change in the atomic state under observation is due to the individuality of the quantum of action. But what in psychology is analogous to the quantum postulate? Remember that the quantum postulate is, for Bohr, merely the empirical discovery of a contingent fact. The world *might conceivably* have been classical and, so it seems, action *might* have been continuous. In psychology, by contrast, the subject and object distinction is a *logical* requirement of description, where the indivisibility of the subject is again a necessary condition for introspective reports of free will, for instance.[56] In fact, the objection is tantamount to a denial that there is any analogy between quantum mechanics and psychology, and therefore no general epistemological lesson is to be learned from quantum mechanics. Still, this way of looking at the situation is an implausible one, I believe.

First of all, if the indivisibility of the quantum of action is merely a contingent fact it is indeed impossible to see how one might reach a logical conclusion about necessary conditions for the acquisition of knowledge on the basis of a simple empirical discovery. The fact is, one might argue, that the description of the quantum of action as a fundamentally discontinuous element makes explicit an internal conceptual tension already existent in the classical framework between particle and wave conceptions of matter. Second, the counterpart analogous to the quantum postulate is indeed Høffding's principle of personality. In psychology the distinction between subject and object is arbitrary whenever a subject attempts to describe its own mental states. In other words, when the content of the mind is part of the mind itself, we cannot easily distinguish between subject and object in the way that an objective description requires.

In Bohr's paper "Biology and Atomic Physics" from 1937 we find the descriptive difficulties encountered in psychology further amplified. After pointing out that, when all is said and done, any report of experience in atomic physics rests on the concepts necessary for all conscious registration of sense impressions, for which the concept of causality is essential, he writes:

> The last remark brings us back into the realm of psychology, where the difficulties presented by the problems of definition and observation in scientific investigations have been clearly recognized *long before* such questions became acute in natural science. Indeed, the impossibility in psychical experience to distinguish between the phenomena themselves and their conscious perception clearly demands a renunciation of a simple causal description on

the model of classical physics, and the very way in which words like "thoughts" and "feelings" are used to describe such experience reminds one most suggestively of the complementarity encountered in atomic physics. I shall not here enter into any further detail but only emphasize that it is just this impossibility of distinguishing, in introspection, sharply between subject and object which provides the necessary latitude for the manifestation of volition.[57]

Here Bohr indirectly pays tribute to Høffding when saying that the difficulties of definition and observation which the physicists now meet in atomic physics had been recognized a long time ago by the psychologists. And here as elsewhere, in using the phrase "problems of definition" Bohr refers to the problems of realizing the ideal of causality in situations where the act of observation has a direct influence on what is observed, as happens in psychology in situations where we cannot clearly separate the awareness of the content of the mind from the content itself. In a similar context in his paper "Light and Life" from 1932, he states this point very clearly:

Indeed, the necessity of considering the interaction between the measuring instruments and the object under investigation in atomic mechanics exhibits a close analogy to the peculiar difficulties in psychological analysis arising from the fact that the mental content is invariably altered when the attention is concentrated on any special feature of it.

And he continues:

It will carry us too far from our subject to enlarge upon this analogy which offers an essential clarification of the psycho-physical parallelism.[58]

This last remark indicates that Bohr not only believed that the subject-object problem arises in quantum mechanics as well as in psychology because of the direct disturbance of the state of the object by observation and reflection on the self, but also thought, like Høffding, that it turns up again in the analysis of the relationship between mind and matter. Thus the thought suggests itself that Bohr regarded kinematic-dynamic complementarity in quantum mechanics as an analogue to what we may call Høffding's introspective-involuntary complementarity in psychology, and that he saw an analogy between wave-particle complementarity and mind-matter complementarity of the sort that comes to expression in Høffding's double-aspect theory. But in the Planck *Festschrift* Bohr also compared the complementarity of the particle-wave picture to Høffding's principle of personality, "the unity of consciousness", a comparison which might have guided him in grasping the complementary aspects of a wave and a particle description.

Recall Høffding's principle of personality. According to this the mind or the consciousness cannot be regarded merely as an aggregate of its various parts or elements, since all elements of the mind or of the consciousness are determined by the mind as a whole as well as the entire mind's being determined by the elements or the parts in virtue of interaction with each other. The elements do have their properties by virtue of being part of conscious life and the mind emanates from the elements. This Høffding called "the law of relation". But since the mind and its elements cannot be separated despite their distinctness there exists an irrational relation, or as Høffding also liked to call it, an antinomy between the continuous life of the mind or consciousness and the discontinuous existence of the elements. Thus this antinomy is linked up with the nature of mind and the unity of personality.[59]

It is precisely this concept of Høffding's which Bohr refers to when he writes:

> In particular, the apparent contrast between the continuous onward flow of associative thinking and the preservation of the unity of the personality exhibits a suggestive analogy with the relation between the wave description of the motions of material particles, governed by the superposition principle, and their indestructible individuality.[60]

Høffding undoubtedly saw conscious thinking as consisting of an involuntary flow of discontinuous items as well of a voluntary synthesis of reflection. So it is quite in harmony with Høffding's antinomy between reflection and involuntary mental life such as sensation and recognition when Bohr in the same paper also writes, "Strictly speaking, the conscious analysis of any concept stands in a relation of exclusion to its immediate application".[61] For according to Høffding such exclusive relations are essential to the nature of mind or mental life.

In several other works Bohr alluded to the epistemological analogy between atomic physics and psychology, just as he also cited examples of a kind similar to those used, as we have noted, by Høffding in his work.[62] Bohr's fondness for Poul Martin Møller's story *En dansk Students Eventyr* (The Adventures of a Danish Student) is well known. In this tale the incompatibility between self-observation and involuntary action is illustrated in the form of the dilemma of The Licentiate, one of the characters in the book. Møller was a professor of philosophy at the University of Copenhagen between 1831 and 1836, and he exerted a powerful influence on Kierkegaard, who at that time was one of his students. His book was Bohr's favorite work of literature as Bohr regarded it as a lesson in epistemology *par excellence*. This is testified to by Rosenfeld, who records that "Every one of those who came into closer contact with Bohr at the Institute, as soon as he showed himself sufficiently proficient in the Danish language, was acquainted with the little book: it was part of his initiation".[63] In another place Rosenfeld tells us, undoubtedly with some exaggeration, that this "delightfully humorous illustration of Hegelian dialectics ... would one day start a train of thought leading to the elucidation of the most fundamental aspects of atomic theory and the renovation of philosophy of science".[64] Although this overstates the case, there is little doubt that the tale brilliantly illustrates some of the problems concerning the separation of subject and object in psychology which seem to have guided Bohr in developing the idea of complementarity.

Bohr himself comments on the tale in the following words: "The author gives a remarkably vivid and suggestive account of the interplay between the various aspects of our position, illuminated by discussions within a circle of students with different characters and divergent attitudes of life".[65] The Licentiate especially is devoted to obscure philosophical speculation to the detriment of his social life. Bohr refers to a scene where a cousin of The Licentiate, The Philistine, blames him for not having made up his mind whether or not to take the practical job that his friends in their kindness have offered him. The Licentiate responds by apologizing, as he explains the difficult situation reflections have put him in. He then says something which Bohr quotes:

> My endless enquiries make it impossible for me to achieve anything. Furthermore, I get to think about my own thoughts of the situation in which I find myself. I even think that I think of it, and divide myself into an infinite retrogressive sequence of "I"s who consider each other. I do not know at which "I" to stop as the actual, and in the moment I stop at one, there is indeed again an "I" which stops at it. I become confused and feel a dizziness as if I were looking down into a bottomless abyss, and my ponderings result finally in a terrible headache.

To this The Philistine replies:

> I cannot in any way help you in sorting your many "I"s. It is quite outside my sphere of action, and I should either be or become as mad as you if I let myself in for your superhuman reveries. My line is to stick to palpable things and walk along the broad highway of common sense; therefore my "I"s never get tangled up.[66]

It is quite obvious that Poul Martin Møller expresses here in poetic form some of the problems of distinguishing between subject and object with respect to the experience of self, by describing "the conditions of analysis and synthesis of so-called psychic experiences" with which Høffding also became preoccupied. Notice further that The Licentiate's problem with the distinction between his many Egos has certain parallels in Høffding's series of alternating objects and subjects.

There are again good reasons to believe that Bohr was thinking of Høffding's defense of the double-aspect theory where he refers to psycho-physical parallelism. As it appears in the paper written for the Planck Jubilee in 1929,

> When considering the contrast between the feeling of free will, which governs the psychic life, and the apparently uninterrupted causal chain of the accompanying physiological processes, *the thought has, indeed, not eluded philosophers that we may be concerned here with an unvisualizable relation of complementarity*.[67]

This statement is in agreement with the opinion Høffding expresses in his last essay on epistemology, which was written around the same time as the Planck Jubilee paper, but it is also in keeping with Høffding's idea that will is superior to both emotion and thought as it saturates all forms of mental activity. A little further on Bohr continues:

> According to the above-mentioned view on the relation between the processes in the brain and the psychical experiences, we must, therefore, be prepared to accept the fact that an attempt to observe the former will bring about an essential alteration in the awareness of volition.[68]

Let us see what Bohr means by such a claim.

Through reading Høffding's book on psychology and through discussions had with him, Bohr had been presented with the double-aspect theory as being the only theory of mind which is both intelligible and in accordance with all empirical facts. Høffding believed that the relation between mind and matter should be characterized as being two aspects of one and the same thing. The existence of both the material and mental features of a person are confirmed by experience, Høffding argued, as well as by the existence of causal connections between physical states and between mental phenomena, but experience cannot verify the existence of a causal connection between the material and mental states of a person. This is due to the fact that it is impossible for the subject to separate sharply the content of its awareness from the awareness itself because part of this content, the mental state, belongs to the knowing subject him/herself, whereas the other part, the physical state, does not.

For these reasons Bohr takes the double-aspect theory for granted. He never puts forward any form of argument like the one above in favor of the theory. What he attempts to do here by invoking the notion of complementarity is not to vindicate a particular theory of the relation between mind and matter, but rather to neutralize a possible objection against the double-aspect theory he had adopted from Høffding. The objection can be put in this way: How is it possible to ascribe two essentially different and sometimes incompatible sets of predicates to one and the same thing, which is what one does in formulating that theory? Bohr's answer is that observation of neuro-physiological processes will influence the awareness of volition so that the two sets of description are not simultaneously applicable. I take this to mean that if we observe certain neuro-physiological processes which are assumed to correspond to certain mental processes such as the making of a decision, the interference with the physical processes in the brain excludes a situation in which the person whose brain processes are being studied can sustain an awareness of the accompanying mental activity. Hence, on the one hand, a subject cannot make any introspective report about the mental enterprise of making a decision if the physical processes in the brain of the subject become the object of scientific investigation. But, on the other hand, neither is it possible to describe in either causal or in quantum mechanical terms the physical processes which might accompany the experience of having a free will.

Bohr rejected the claim that the problem of free will might be explained by appealing to indeterminism at the atomic level as some physicists have thought. As he said in 1932,

> In fact, according to the parallelism, the freedom of the will is to be considered as a feature of conscious life which corresponds to functions of the organism that not only evade a causal mechanical description but resist even a physical analysis carried to the extent required for an unambiguous application of the statistical laws of atomic mechanics.[69]

So what Bohr claimed is that the disturbance analogy not only applies to the introspective analysis of self-conscious reflection and involuntary experience, but may also apply to particular investigations of the relation between mental states and physiological states. It is, indeed, the existence of the disturbance of the object of investigation which guarantees that descriptions of mental states and neuro-physiological states are complementary and not contradictory, since the influence of the last on the first creates an observational situation in which a simultaneous description of the phenomenon as both mental and physical is impossible.

Indeed, it might be objected that Bohr cannot have taken over the concept of psycho-physical parallelism from Høffding, since the latter calls his theory of mind and matter the identity theory.[70] But to my mind such a claim does not prove anything. Høffding knew, of course, that his position has sometimes been called parallelism, but he argued that this label was inadequate if taken in a literal sense since, given this sense, the mental and the material are considered as two isolated series of states, like two rails.[71] Such a dualist point of view wasn't Høffding's; his view was monistic, and so was Bohr's despite his talk of psycho-physical parallelism.

Evidently, Bohr's general point seems to have been that the difficulty of distinguishing between subject and object in the domain of psychology is to be solved by taking into account that causal and intentional descriptions are *complementary*, in the sense that experiences under different observational and introspective conditions mutually exclude each other, but together they are exhaustive in describing the total conscious and physical life of a human being.

The complementarity view thus harmonizes both with Høffding's conception of an antinomy between the causal description of volition and the experience of free will which he believes characterizes mental life as well as with his theory of the antinomy between the mental and the physical. In fact, Bohr echoed the result of the discussions between Høffding and himself during 1928–1929 when in the second essay from 1929 he wrote:

> Hoping that I do not expose myself to the misunderstanding that it is my intention to introduce a *mysticism* which is incompatible with the spirit of natural science, I may perhaps in this connection remind you of the peculiar parallelism between the renewed discussion of the principle of causality and the discussion of a free will which has persisted from earliest times. Just as freedom of the will is an experiential category of our psychic life, causality may be considered as a mode of perception by which we reduce our sense perceptions to order. At the same time, however, we are concerned with idealizations whose natural limitations are open to investigation and which depend upon one another in the sense that the feeling of volition and the demand for causality are equally indispensable elements in the relation between subject and object which forms the core of the problem of knowledge.[72]

The ordinary causal description of classical mechanics has to be abandoned in atomic physics and replaced by a probabilistic description. Nevertheless, according to Bohr, the causal relationship remains the form of perception in that it is this which orders our sensory experience just as the freedom of the will is the form of our experience of our mental life in spite of the fact that this purported freedom may be challenged. The feeling of freedom is essential to the mind, but whether or not this feeling is grounded in reality is an open question which empirical methods, at any rate, are unable to settle. It is the "impossibility of distinguishing, in introspection, sharply between subject and object which provides the necessary latitude for the manifestation of volition,"[73] and it is in virtue of this impossibility that the question remains open.

Hence, Bohr considered the causal relationship and freedom of the will to be idealizations, although the concept of free will is necessary for the characterization of a subject in the same way as the concept of causality is logically indispensable for the specification of an object.

We have seen, then, that Bohr constantly makes comparisons between the new situation in quantum mechanics and the situation in psychology, with which Høffding had acquainted him, in order to elucidate and explain what he means by complementarity, because from the very start he seems to have been cognizant of the presence of an epistemological analogy between these two fields. Høffding had taught him that the methodology of psychology rested on a necessary combination of a rational, introspective mode of description and an empirical, naturalistic mode of description imposed on us by the aim psychology sets itself: to give an account of the relation between the mind experienced as a whole and its various elements. Consequently, it was this methodological dualism which reappeared in Bohr's understanding of quantum mechanics as it was a formative factor in the development of the complementary framework for the description of atomic processes. However, Høffding had not only applied his theory of knowledge and scientific methodology to psychology but also to biology and sociology. So it is not surprising to see Bohr also making an excursion to these fields in order to find similar complementary aspects of descriptions as within the atomic world.

3. BOHR ON BIOLOGY

In the Introduction we saw that in the last interview he gave Bohr recalled that, as a student, he was preoccupied not only with the problem of description in psychology but also with the similar situation which obtained in biology. This interest in biology and the description of living organisms was already implanted in Bohr's mind, dating from the time when he listened in on the discussions among his father, Høffding, Christiansen, and Vilhelm Thomsen on the controversy between mechanism and vitalism. Bohr himself mentioned these discussions in his 1955 paper "Physical Science and the Problem of Life". After quoting a fairly long passage from one of his father's works, in which Christian Bohr pointed out that alongside mechanical descriptions, teleological considerations were indispensable in the study of living organisms, he says:

> I have quoted these remarks which express the attitude in the circle in which I grew up and to whose discussions I listened in my youth, because they offer a suitable starting point for the investigation of the place of living organisms in the description of nature.[74]

As Høffding was a prominent member of the group, his opinions doubtless played an important role and his great familiarity with Kant's writings is demonstrated by the way in which he and Bohr's father acknowledge the heuristic value of teleological considerations in addition to purely mechanical considerations in biology.

In Bohr's thought we find a point of view very close to that of Kant, Høffding and his father, and there is little doubt that the way Høffding had analyzed the problems of description in biology may have influenced Bohr in his formulation of complementarity with respect to biology. Nevertheless, it was not until shortly after Høffding's death that Bohr for the first time wrote about the application of complementarity to biology in an address called "Light and Life" at the International Congress of Light Therapy in Copenhagen in August 1932. Here Bohr did not consider the notion of complementarity to apply exclusively to quantum mechanics, but to be a principle applicable to any epistemological situation in which Høffding's criterion for reality cannot be used. In this paper as well as in "Biology and Atomic Physics" from 1937 he attempted to show the fruitfulness of using the principle of complementarity in the description of living organisms. In 1932 he outlined the situation as follows:

> ... if we were able to push the analysis of the mechanism of living organisms as far as that of atomic phenomena, we should scarcely find any features differing from the properties of inorganic matter ... however ... the conditions holding for biological and physical researches are not directly comparable, since the necessity of keeping the object of investigation alive imposes a restriction on the former which finds no counterpart in the latter ... On this view, *the existence of life must be considered as an elementary fact that cannot be explained*, but must be taken as a starting point in biology, in a similar way as the quantum of action, which appears as an irrational element from the point of view of classical mechanical physics, taken together with the existence of elementary particles forms the foundation of atomic physics. The asserted impossibility of a physical or chemical explanation of the function peculiar to life would in this sense be analogous to the insufficiency of the mechanical analysis for the understanding of the stability of atoms.[75]

Much of what Bohr says here about biology sounds like an echo of Høffding.

Høffding too believed that life is an unexplainable empirical fact which cannot be defined in terms of physics or chemistry. Likewise Høffding regarded everything which is not subject to a causal explanation as representing that which is beyond the bounds of intelligibility. Thus, accordingly, Høffding's argument for the inexplicability of life may be understood as follows: life is a feature which characterizes organisms as indivisible wholes, and the relation between a whole and the sum of its parts cannot be accounted for in causal terms. That is, the analysis of wholeness is irreducible to an analysis of the causal interrelations among the components of a system. From the point of view of immediate experience living organisms possess their own intrinsic "individuality". In other words, there exists an antinomy between the organism apprehended in experience as a whole and the organism characterized by a causal mode of description of the relations among its parts which a rational understanding demands in order to give an account of the functions of a living being; so consequently, life represents what is beyond the bounds of intelligibility. But from this it does not follow, Høffding would say, that a thorough understanding of the vital structures and functions of biological organisms can be achieved on premisses other than physico-chemical ones. Vitalism, as an ontological thesis about the presence of some non-physical entity within living organisms, has no place in a coherent conception of nature; but neither does animate nature support a purely mechanistic point of view.

A similar line of reasoning seems to underlie the quotation from Bohr. He looks upon life as an irreducible empirical fact analogous to that of the quantum of action. This point of view is repeated in 1937:

> ... the existence of life itself should be considered, both as regards its definition and observation, as a basic postulate of biology, not susceptible of further analysis, in the same way as the existence of the quantum of action, together with the ultimate atomicity of matter, forms the elementary basis of atomic physics.[76]

In fact, the content of the two papers is very much the same. On the one hand, Bohr believes that "we all agree with Newton that the ultimate basis of science is the expectation that nature will exhibit the same effects under the same conditions". Much progress in physiology and biology has been reached by using chemical and mechanical models in explaining the internal as well as the external reactions of the organism. Even if we were able to aim at an analysis of the biological mechanisms at the atomic level, we would not be confronted with physical features other than those of inorganic matter. There is no reason to expect that we will find any law of biology which is at variance with physical or chemical laws. Thus, vitalism cannot be supplied with an unambiguous foundation. On the other hand, life is an irrational element from the point of view of physics just as the quantum of action is in relation to classical mechanics. That is, life is an essential feature of animate matter which is eliminated by any attempt at describing the physical processes taking place in the parts of an organism. Life must be sustained during examination if the object of investigation is to be described as a biological object and not merely as a physical object.

However, if the property of life is retained, we discover that biological organisms display such holistic characteristics as self-preservation and reproduction that in contexts where these holistic characteristics play a part they are inaccessible to an unambiguous causal mode of description, even though a far-reaching understanding of the chemical and physical aspects of many typical biological reactions is something we are in possession of. Both living organisms and the quantum of action are characterized by a non-causal feature of wholeness. Thus Bohr writes:

> Indeed, the essential characteristics of living beings must be sought in a peculiar organization in which features that may be analyzed by usual mechanics are interwoven with typically atomistic features to an extent unparalleled in inanimate matter.[77]

Certainly, here, "atomistic features" does not refer to certain properties of atomic systems but to traits of individuality or wholeness which are also seen in the quantum of action. Bohr describes organisms as possessing an organization which must be maintained in order for them to stay alive. He thus feels that not all teleological considerations can be replaced by mechanistic considerations as the vital functions proper are unanalyzable in terms of physical and chemical descriptions. As he says, "The concept of purpose, which is foreign to mechanical analysis, finds a certain field of application in biology".[78] That is, the experience of living organisms as wholes demands the use of finalistic concepts in the description of their behavior. So the point Bohr wants to make is about how to *describe* the objects of biological science, not about what they ultimately *are*. He takes it for granted they are purely physical systems, but they are ones which manifest the properties of life.

We have, apparently, two alternative sets of descriptions in biology, both of which are indispensable in the characterization of organic matter. The resolution of the conflict between the two levels of description is articulated by Bohr as follows:

> In fact, we are led to conceive the proper biological regularities as representing laws of nature complementary to those appropriate to the account of the properties of inanimate bodies, in analogy with the complementary relationship between the stability properties of the atoms themselves and such behavior of their constituent particles as allows of a description in terms of space-time coordination.[79]

At the same time, it is important for Bohr to stress that such a view lies equally distant from vitalism and from mechanism, there being no question of any attempt to introduce specific biological laws in conflict with well founded physical and chemical rules. And this depends, he writes, on the fact that

> the possibility of avoiding any such inconsistency within the frame of complementarity is given by the very fact that no result of biological investigation can be unambiguously described otherwise than in terms of physics and chemistry, just as any account of experience even in atomic physics must ultimately rest on the use of the concepts indispensable for a conscious recording of sense impressions.[80]

That is, in both biology and atomic physics phenomena cannot be described unambiguously in terms of concepts other than the classical ones of continuity despite the property of life and the element of the quantum of action, for these concepts are the only ones which supply us with an objective account of reality. On the contrary, teleological concepts have no explanatory power as such since "any scientific explanation necessarily must consist in reducing the description of more complex phenomena to that of simple ones"; they refer only to the immediate experience as regulative principles. Bohr felt, nevertheless, that teleological considerations are necessary for the characterization of living organisms provided these considerations do not conflict with physical and chemical laws, and this is possible only if experience that demands finalistic and physical descriptions respectively is acquired in incompatible situations.

This is in fact what Bohr believed to be the case. Since the presuppositions underlying teleological and mechanistic descriptions respectively are mutually exclusive the two types of description cannot be simultaneously sustained and thus do not come into conflict with each other. Thus, Bohr held that the finalistic mode of description is complementary to the physical mode of description in virtue of the fact that the observational conditions required for each, taken individually, are mutually incompatible. Or, as one might also put it, since the property of life applies to the object taken as a whole whereas a mechanistic description applies merely to the object considered as consisting of the sum of its parts, mechanistic and teleological descriptions are mutually exclusive and yet may be deployed complementarily.

However, Folse has pointed out that in his earliest essays on biology Bohr seems to have disregarded the fact that in quantum physics the impossibility of separating the states of two interacting systems is a physical consequence of the quantum postulate and saw the impossibility in biology of describing the vital functions of a certain organism in physical terms and keeping that organism alive as an experimental problem which might be overcome by refined techniques. Consequently, Folse argues, Bohr did not recognize until much later that when recourse is had to complementary descriptions, it is because it is open to us to describe biological phenomena on two distinct levels.[81] If this claim is true it looks very much as if in the thirties Bohr was prey to the same ambivalence we observed in Høffding, in the beginning, as to the impossibility of an exhaustive physical description of living things.

Folse is correct insofar that in his early essays Bohr does not directly endorse what he explicitly states in his 1955 article, namely, that

> The basis for the complementary mode of description in biology is not connected with the problems of controlling the interaction between object and the measurement tool, already taken into account in chemical kinetics, but with the practically inexhaustible complexity of the organism.[82]

In fact Bohr begins his discussion of the problem of life in "Light and Life" by noting that the quantum of action brings out the dimension of wholeness inasmuch as the atomic system cannot be distinguished from the means of observation.[83] The inseparability of the phenomenon under observation and the action of the measuring instrument is what prohibits a completely mechanistic description of the atomic system. Instead the observing and observed systems form an interacting whole with respect to which we can only draw an arbitrary division between subject and object. It is this fact which, according to Bohr, necessitates the complementaristic account in quantum physics. One therefore naturally gets the impression that Bohr actually saw a resemblance here between atomic physics and biology in the sense that he believed there was a similar argument in favor of complementary descriptions in biology.

However, the fact of the matter is quite the opposite. There is no such argument. An interaction with the functions of the biological organism under investigation is a precondition for a physical analysis of these processes. It is not the case that the observing and observed system form an interacting whole which is indivisible. As was already said, the necessity of sustaining life in the phenomenon under observation implies a significant limitation on the possibilities of the investigation of those bio-chemical functions which provide the empirical basis for a mechanistic analysis. Both in 1932 and 1937 Bohr compared this limitation to the situation in quantum mechanics, in which the stability of an atomic system, owing to the quantum postulate, sets limits to the kinds of systematic analysis possible. But contrary to what happens in the case of the quantum of action when in interaction with the atomic system, the property of life ceases to be, given radical intervention in the workings of the vital organs. Hence, it seems as if it might rather be a technical problem whether or not an organism stays alive in the course of an investigation of its life processes.

In his later writings Bohr is more explicit in holding that the impossibility of keeping a certain organism alive in the course of a process of investigation is rather a matter of empirical fact, although he still misleads his readers by saying that the complementary mode of descriptions in biology is due to "the practically in exhaustible complexity of the organism". But now he focuses more on a second argument, according to which the observational conditions necessary for a purely mechanistic definition of the vital functions are incompatible with those necessary for the manifestation of life. The first set of conditions is such as to require that the states of an organism be defined in isolation from its interaction with its environment for such interactions to be described in mechanistic terms, while, according to the second set, the interaction of the organism with its environment is essential to the manifestation of the properties to which the term "life" refers. However, this argument also occurs in the earlier essays on biology, as Folse himself notes. As Bohr wrote in 1937, "The only way to reconcile the laws of physics with the concepts suited for a description of the phenomena of life is to examine the essential difference in the conditions of the observation of physical and biological phenomena".[84] That is, on the one hand, in order for the functions of living organisms to be described causally a sharp separation of the organic system and the environment is required, so that both organism and environment can be subdivided into their various components. On the other hand, "The incessant exchange of matter which is inseparably connected with life will even imply the impossibility of regarding an organism as a well-defined system of material particles like the systems considered in any account of the ordinary physical and chemical properties of matter".[85] In the case where the organism is interacting with its surroundings, it is considered as a whole. Thus Bohr's reasoning is then that since the observational conditions required for describing the organism in isolation and those for describing its interaction with its surroundings are incompatible but both necessary for any comprehensive account of a living organism, the two modes of description are complementary. But, even though Bohr did not mention the subject-object distinction himself when discussing biology, he would probably have agreed with Høffding that the characterization of an organism as a totality gives rise to a problem concerning the separability of subject and object in biology as it does in psychology, because whenever we look upon an organism as a whole the concept of causality from part to whole fails to apply and the resulting teleological description is therefore not objective, i.e. unambiguous.

Neither Høffding nor Bohr took a positivistic attitude to the old mechanist-vitalist debate. They did not see the debate as one concerning a pseudo-problem as did the positivists, who abandoned all metaphysical talk, deeming it senseless.[86] For although they agreed with the positivists that science describes a single natural realm and so rejected with them the ontological claims of vitalism, of entities different from material entities, they both endorsed the epistemological motives behind vitalism. Both held that the difference between inorganic matter and living organisms does not consist in the substance of which they are composed and therefore that they are not governed by inconsistent laws. But where the positivists attempted to do away with teleological explanations by reducing teleological statements to physicalistic statements, Høffding and Bohr argued that life itself cannot be described in terms of causal statements for the simple reason that the teleological language relating to the life functions cannot be "translated" into physicalistic statements because it referred to properties which "supervene" on the member organs of the organism and are not reducible to them. For them teleological language expresses fundamental features of our common experience of living organisms which cannot be captured by a mechanistic language. Thus, teleological descriptions cannot be translated into or replaced by mechanistic ones: each must be understood as making reference to organisms observed in situations that are such as to make the description of one complementary to that of the other.

Chapter VII

1. THE OBJECTIVITY OF KNOWLEDGE

The concept of continuity was as essential to Bohr's theory of knowledge as it was to Høffding's in virtue of its being for both a precondition for the acquisition of objective empirical knowledge. Thus it is the concept of causation along with the concepts of space and time which in physics enables us to distinguish between subject and object in the experience given to us through the senses, and the concept of causation alone, or perhaps together with the concept of time, which in psychology makes it possible, when it is, to distinguish between the subject of awareness on the one hand and the experience, thoughts and emotions constituting the content of consciousness on the other. However, Bohr eventually realized that the quantum of action created a problem for the viability of causal descriptions of individual atomic objects required by the ideal of causation, inasmuch as the causal mode of description had to be considered complementary to the spatio-temporal mode of description. What became clear to Bohr was that the difficulty of applying the causal mode of description to the domain of quantum mechanics is epistemically equivalent to difficulties of a similar kind confronting Høffding in psychology and biology, where the subject-object distinction is blurred, owing to the character of wholeness and unity, which is distinctive for the subject's experience of phenomena. Høffding saw the difficulties in psychology and biology as resulting from an unavoidable antinomy between the use of rational and empirical descriptions, that is, between the application of causal and holistic descriptions respectively, while Bohr perceived the solutions to the difficulties to lie in the use of complementary descriptions in quantum mechanics, as well as in psychology and biology.

The inevitable interaction between the object and the measuring instrument in quantum mechanics has a holistic aspect comparable to that encountered in psychology and biology in cases where subject interacts with object. The aspect of integrity is apprehended as such because no sharp distinction can be made between subject and object; it appears, that is, because it is impossible to describe the process of observation of the object in causal terms. The reality of the object independent of our observations is therefore, according to Bohr, called into question: the quantum of action throws "new light upon the old philosophical problem of the objective existence of phenomena independently of our observations".[1] This and other statements made by Bohr in the Como paper and the two 1929 papers have been interpreted by some philosophers to mean that Bohr held either a subjectivist or a microphenomenalistic attitude towards atomic objects.[2] Such a reading might find some support in a passage like the following:

> We have learned from the theory of relativity that the expediency of the sharp separation of space and time, required by our senses, depends merely upon the fact that the velocities commonly occurring are small compared with the velocity of light. Similarly, we may say that Planck's discovery has led us to recognize that the adequacy of our whole customary attitude, which is characterized by the demand for causality, depends, solely upon the smallness of the quantum of action in comparison with the actions with which we are concerned in ordinary phenomena. While the theory of relativity reminds us of the subjective character of all physical phenomena, a character which depends essentially upon the state of motion of the observer, so does the linkage of the atomic phenomena and their observation, elucidated by the quantum theory, compel us to exercise a caution in the use of our means of expression similar to that necessary in psychological problems where we continually come upon the difficulty of demarcating the objective content.[3]

Høffding too had already in 1921, as we have seen, claimed that the theory of relativity yields a subjective perspective on all phenomena, inasmuch as properties such as space and time which were previously ascribed to an object in an absolute sense in Newton's theory can in fact only be specified in relation to a knowing subject. So there is good evidence for the belief that Bohr's observation owes something to their many discussions. The claim, however, did not mean that for Høffding objective knowledge of the physical world is impossible or that we do not observe objects themselves. And, as we shall see, there is no reason to assume that Bohr thought differently than Høffding.

What Høffding had in mind when making this claim was something like this: The theory of relativity exemplifies in the fullest possible way the general epistemological requirement that we take the knowing subject into account when justifying any claim to possession of objective empirical knowledge. The fact is, according to him, that the theory of relativity has it that spatio-temporal attributes are objective but not absolute or inherent in the sense that they only characterize an object relative to a frame of reference which is selected by the observer.

If the description of the content of our experience is to be an objective one it is required that its relationship to the subject can be specified. The content of experience must be described in terms of continuity if it is to satisfy the requirement for what it is for an experience to be concerned with a physical object that is independent of the experiencing subject. The ascertainability of continuous connections serves as the criterion of whether a particular phenomenon is real or not. Since the applicability of the concept of continuity to subjective experience is the criterion of that which yields the objective existence of the content of the experience, that is, of the independent existence of the observed phenomenon, Høffding inferred that we can only specify a thing, an immediately given whole, in virtue of its relations to other things. A physical object is thus a phenomenon that is immediately apprehended as a whole and which is brought into continuous relations with other phenomena also apprehended as wholes. That is, so long as a phenomenon exists only as an immediately given integral experience, it cannot be regarded as an independent object, even though the phenomenon, as an observed object, is presented to, not created by, consciousness. The phenomenon is an object existing independently of the observation only if it can be brought into causal relations with other phenomena. Høffding had once defined the properties of a thing as the ways in which it is influenced by or influences other things. Things or physical objects – whether macroscopic or microscopic – cannot merely only be known as objects but can also only be defined as such in relation to other things. But if a physical object can only be known and defined as an object on the strength of its relations to other physical objects, it becomes meaningless to ascribe to this object properties independent of these relations; consequently, a physical object cannot unambiguously be attributed absolute properties. One might say that there is no room for a notion of absolute, intrinsic properties in Høffding's theory of knowledge because the immediately given phenomenon cannot in itself be characterized as an independent object; this is possible only after its relation to the subject has been determined through a complete causal description of the phenomenon. So properties are always contingent upon our cognitive capacity.

Thus, Høffding believed that the theory of relativity supports a view to the effect that the subject, via its forms of thought, is responsible for the forms of appearance of an object by showing that the notion of inherent or absolute spatio-temporal properties central to Newton's theory are abstractions and idealizations. The epistemological foundation of Newton's theory is in fact the assumption, Høffding would say, that both the content of perception, the phenomena, as well as the forms of perception are given to us in a passive act of cognition by which the perceiving subject creates ideas and sense impressions corresponding to the inherent and absolute properties of the object. These ideas and impressions are therefore assumed to represent an independent reality behind the phenomena. However, the epistemological foundation of the theory of relativity is, by contrast, the assumption that the subject is active in virtue of the forms of thought it possesses, in terms of which the phenomena are perceived and which are preconditions of the possibility of sensory experience. There is no way in which the forms of thought can both represent a world behind the phenomena and at the same time be necessary for there to be any such thing as sensory experience at all. What Høffding therefore meant by his reference to the subject's perspective was that the theory of relativity confirms the thesis that all the properties we ascribe to an object are only attributable to it in virtue of the forms of thought possessed by the subject. The ascription of values of space and time as well as of momentum and energy to the object vary according to the observer's choice of reference system. However, the properties are still objective, albeit relational, because when they are ascribed to the object its behavior is thereby rendered subject to a causal space-time mode of description.

This is probably also what Bohr had in mind by claiming "that the theory of relativity reminds us of the subjective character of all physical phenomena". That this is what he meant can be seen from the fact that Bohr did not think "that the subjective character of all physical phenomena" is a threat to "the objective content" of experience. The theory of relativity satisfies the criterion of objective empirical knowledge by describing phenomena causally connected in space and time. The forms of perception had, throughout the history of science, guaranteed that "the co-ordination of our experience of the external world" is objective. "Yet occasionally just this 'objectivity' of physical observations becomes particularly suited to emphasize the subjective character of all experience".[4] It is, however, quantum mechanics and psychology which challenge that criterion of objective empirical knowledge. That interpretation is also endorsed by Bohr's statement in the other 1929 paper: Although "the theory of relativity which, by a profound analysis of the problem of observation, was destined to reveal the subjective character of all the concepts of classical physics", it "approaches, in a particularly high degree, the classical ideal of unity and causality", which means that "the conception of the objective reality of the phenomena open to observation is still rigidly maintained".[5] But the classical ideal of objectivity is not attainable in the description of atomic phenomena.

As we have seen, Bohr believed that the discovery of the quantum of action had brought to an end the sharp separation between object and instrument since it implies that the exact simultaneous position and momentum of an object cannot be measured. But Bohr also held that an object cannot meaningfully be said to possess these exact values simultaneously. I shall, following Murdoch, call the latter claim Bohr's *indefinability thesis*.[6] As Bohr put it in the late twenties, referring to the unpredictability of the future course of the atomic object owing to the indivisibility of the quantum of action:

> Obviously, these facts not only set a limit to the *extent* of the information obtainable by measurements, but they also set a limit to the *meaning* which we may attribute to such information. We meet here in a new light the old truth that in our description of nature the purpose is not to disclose the real essence of the phenomena but only to track down, so far as it is possible, relations between the manifold aspects of our experience.[7]

However, various suggestions have been put forward to explain why Bohr held this position. Some philosophers, such as Adolf Grünbaum and Paul Feyerabend, have dismissed the claim that Bohr's reasons for the indefinability thesis stem from a general philosophical doctrine. Instead they both maintain that his grounds are based on ontic arguments.[8] Contrary to what they claim I shall argue that Bohr's reasons for holding the indefinability thesis up to 1935 is based, first and foremost, on the same epistemic arguments as those which were central to Høffding's philosophy. The problem, however, when discussing

Bohr's philosophy of physics is that most philosophers have read Bohr as if the arguments for his philosophical position had remained the same from 1927 to his death and have failed to note the changes in his terminology as well as those that he made with respect to some of the central arguments after 1935.

In classical mechanics it is possible to give an account of the properties of an object in the form of a causal space-time description of the state of a system independent of observational interaction. The theory of relativity does not challenge this in spite of the fact that it yields no place in the system for inherent and absolute states. But such an account is blocked in quantum mechanics because of the feature of wholeness characterizing the interaction between the object and the measuring instrument.

> The limit, which nature herself has thus imposed upon us, of the possibility of speaking about phenomena as existing objectively finds its expression, as far as we can judge, just in the formulation of quantum mechanics.[9]

The limit with respect to the description of atomic phenomena, which Bohr is talking about here, is that these objects cannot be ascribed properties independently of the experimental set-ups in which they make themselves known. Bohr formulated his view in the Como paper as follows: "It must be kept in mind that ... radiation in free space as well as isolated material particles are abstractions, their properties on the quantum theory being definable and observable only through their interaction with other systems".[10] Bohr thus emphasized that concepts such as position and momentum in quantum mechanics are proved to be relative concepts, being only applicable in relation to certain experimental set-ups because of the quantum of action, and not applicable to free isolated and unobserved objects. However, this additionally implies – inasmuch as one cannot make a causal mode of description of atomic particles – "the impossibility of a strict separation of phenomena and means of observation".[11] Quite in agreement with Høffding, Bohr determined things as "real" only if they figure in a causal connection.

So it would seem that we find a similarly strong empiricist attitude in Bohr towards atomic properties as we saw taken by Høffding. Referring to the indispensability of classical concepts in the description of atomic phenomena he says, for instance:

> it is equally important to understand that just this circumstance implies that no result of an experiment concerning a phenomenon which, in principle, lies outside the range of classical physics can be interpreted as giving information about independent properties of the objects, but is inherently connected with a definite situation in the description of which measuring instruments interacting with the objects also enter essentially.[12]

Quantum mechanics confirms the epistemological lesson which the theory of relativity has taught us, to the effect that the causal space-time mode of description of experience does not support an ascription of inherent properties to the object. Classical concepts are abstractions and idealizations if the forms of perception, as in classical mechanics, are thought of as mental constructions representing inherent properties of the object independent of observation. By contrast the theory of relativity shows that properties described by the classical concepts are not absolute but are relative to a frame of reference selected by a subject. Quantum mechanics, however, goes one step further. For measurement, according to Bohr, not only cannot "be interpreted as giving information of independent properties of the object", but properties on the quantum theory are definable and observable only through an interaction of the object with other systems in such a way that we cannot define its future behavior on the basis of observation of its initial state.

> The elucidation of the paradoxes of atomic physics has disclosed the fact that the unavoidable interaction between the objects and the measuring instruments sets an absolute limit to the possibility of speaking of a behavior of atomic objects which is independent of the means of observation.[13]

The measurement of a well-defined value of one of the observables providing the initial state of the system excludes the measurement of a well-defined value of the other. This means that the properties of an atomic object can only be understood in relation to the measuring instrument.

Of course, one might retort that sometimes it is possible to ascribe independent properties to an object: namely, if we measure the position of the object twice and know the size of the interval of time between the two measurements, we may calculate the average velocity of the object. Bohr admits this but adds that such a property ascription is a mere idealization.

> Indeed, the position of an individual at two given moments can be measured with any desired degree of accuracy; but if, from such measurements, we would calculate the velocity of the individual in the ordinary way, it must be clearly realized that we are dealing with an abstraction, from which no unambiguous information concerning the previous or future behavior of the individual can be obtained.[14]

Such a velocity is an abstraction because it cannot be determined on the basis of measurement, and hence its calculation, being empirically unjustifiable, cannot be used to predict both the prior course of the object before the first measurement of the position was taken as well as the course subsequent to the second measurement. Thus, it is no longer a question of ascribing properties to the atomic phenomena independently of certain experimental circumstances that determine the conditions of observation and definition.[15]

Bohr's reason for denying that an object possesses a well-defined momentum and position is this: since it is impossible for the simultaneous values of both observables necessary for a description of the behavior of an independent object to be determined on the basis of information gained by measurement, simply because the act of measurement *disturbs* the atomic object, there would be nothing on which to base the ascription of properties to the state of the system at any given time when they are conceived of as being truly ascribed independently of the action of the measuring instrument. Only that which can be experienced can be defined unambiguously, that is, the atomic object possesses well-defined properties only with respect to experiential conditions. The ascription of a property is warranted by observation, and if none can be warranted by observation, there are no more epistemic grounds for defining one independently of it. Descriptions of results reached in differing observational situations will therefore be complementary – jointly they will yield an unambiguous and exhaustive account of the experienced phenomena.

The structure of the entire argument underlying Bohr's complementarity thesis may be summarized in the following eight steps:

(1) That it be possible to produce a causal space-time description of our perceptions constitutes the criterion of reality for them.
(2) The criterion of reality allows us to distinguish between the knowing subject, who formulates the description, and the known object that is so described.
(3) The quantum of action involves an uncontrollable disturbance of the atomic system brought about by the measuring instrument, so it is impossible, at one and the same time, to measure with precision its momentum and position, rendering it unamenable to a causal spatio-temporal description.
(4) Therefore, it is not possible to distinguish the knowledge of the atomic system from that of the measuring instrument.
(5) This feature of inextricability entails that we are not epistemically justified in ascribing reality to the atomic system independently of our ability to observe it.
(6) Therefore, the atomic system does not possess any properties independently of the observational situation on the basis of which the values of these properties are known.
(7) If such properties can be attributed to the atomic system just when there are observational circumstances which are mutually exclusive, and if they are represented by canonically conjugate variables, then these properties are complementary.
(8) Hence, since a complementary property can be ascribed reality just if it can be observed, it has no cognitive meaning to define a precise value for the other one which cannot be observed at the time.

It has no meaning, one may say, because in the absence of any interaction of measurement, the objective relation to which the concept refers ceases to exist. There is nothing for it to refer to.

This line of reasoning constitutes the main argument from epistemic considerations behind Bohr's defence of the indefinability thesis around 1930. Bohr's assistant at that time, Rosenfeld, once confirmed that Bohr's view was that the state of an atomic system should be considered as being that of the relations obtaining between the system and the experimental set-up instead of that of inherent and independent properties.[16] The same point of view is taken by Feyerabend. He says "that complementarity asserts the relational character not only of probability, but of all dynamical magnitudes".[17]

One might ask, of course, whether Bohr believed that the relational properties of the object were created by the process of observation. As we have seen, he argued in the late twenties and early thirties that the interaction between the object and the measuring instrument disturbed the observed phenomena. If this is true, the value obtained through measurement might be said to be created by the instrument, or is perhaps to be thought of as a disturbed value resulting from an interference with a pre-existing value. There is, however, very little which seems to confirm such a view. Bohr certainly believed that the measured value was part of a phenomenon immediately given in observation, as did Høffding with respect to all phenomena. The value of the measurement of what is observed is neither created by the process of observation nor is it a result of a measurement taken at the point when the disturbing interaction ceases taking place. The atomic phenomenon is known in virtue of the relation of a measured value to the measuring instrument and is a phenomenon which is experienced as an immediately given whole. If the measured property were a created value the phenomenal object would not appear as an immediately given whole imposing itself upon the subject, or if it were a result of a disturbed pre-existing value, it would imply the existence of another sphere of reality behind the experienced phenomenon. However, Bohr neither believed in the subjectivity of all experience – he merely talked about the subjective character of all physical experience – nor did he hold the existence of a transphenomenal world as part of his view. Instead of holding that both the content of experience as well as the forms of experience are determined by the subject, which idealists and phenomenalists would do, or holding that they are determined by the object, as realists would do, he maintained with Kant and Høffding that the content is given as part of the object but that the forms are given as part of the subject.

In this context it is important to make clear that even though Bohr, on a couple of occasions, stated that all doubt about the reality of atoms had been swept away, he at the same time emphasized that this was on the assumption that their reality was contingent upon their being observed, and that it makes no sense to speak of their reality independent of observation.[18] As long as we are able to describe the observed object and its properties in causal and spatio-temporal terms we are justified in talking about the object as existing objectively, independently of the knowing subject. But, since the interaction between the atomic phenomena and the measuring instrument cannot be controlled owing to the individuality or indivisibility of the quantum of action, it is impossible to give a full account of the dynamical and kinematical properties of the object, and it is therefore impossible to distinguish sharply between its causal behavior and the causal behavior of the means of observation. There are, furthermore, many properties of the atomic object other than the kinematic and dynamical properties, such as mass, parity and electric charge, which can be ascribed to the object only on the basis of a causal description of how the measuring instrument works. So by stating that "the often expressed skepticism with regard to the reality of atoms was exaggerated" Bohr's intention was to say that atoms have been proved to exist in so far as we possess objective knowledge of them – which we clearly do as it has been possible to explain many experimental results as effects of processes involving atomic objects. Bohr did not want to be associated with the doctrine that atoms are merely fictions or heuristic constructions, introduced in order to organize experience in the most useful way. The concept of the atom does indeed serve the purpose of synthesizing our experience in the most appropriate way, but also makes reference to something real if, by attributing to atoms causally effective properties, we are enabled to describe our experience in an unambiguous way. But since the properties most essential to the characterization of atomic objects as real cannot be defined simultaneously, "words like 'to be' and 'to know' lose their unambiguous meaning"[19] if they are used to refer to objects considered in isolation, whose properties are not determined in relation to a particular observational situation.

The epistemic argument for the indefinability thesis is only superficially like another argument which is widely held to be endorsed by Bohr. According to this argument Bohr's statements are given a positivistic reading.[20] It is held that the indefinability thesis rests upon a positivistic theory of meaning according to which a sentence that ascribes a property to an object is cognitively meaningful if, and only if, it is possible by means of observation to verify it or determine its truth value. Hence, since a sentence which attributes at one and the same time both a definite momentum and a definite position to an atomic particle cannot be confirmed by sensory experience, it has no cognitive content through having no descriptive content. There is clear evidence, I shall argue, that Bohr was later to subscribe in part to this component of the positivist doctrine of meaning. But in spite of the strong empirical element in his epistemology he rejected the other component of it, according to which the conditions constituting the verification or falsification of such a sentence are to be characterized in purely sensory terms.

Even though Bohr never spoke about truth and the kind of concept it is, he endorsed, I think, the same concept of truth as did Høffding, holding that a sentence which ascribes simultaneously an exact momentum and position to an atomic object is true only if it can be coherently connected with other true sentences which jointly describe our sensory experience and all of which can be derived from a consistent theory. This interpretation of Bohr's understanding of truth explains what he meant by saying that "the possibility of an unambiguous use of classical concepts solely depends upon the self-consistency of the classical theories from which they are derived". However, a sentence predicating classical state parameters to an atomic object is inconsistent with the theoretical framework of quantum mechanics, including Heisenberg's uncertainty relations, which successfully describe our experience of the atom, and so such a sentence cannot be assigned a truth value nor, consequently, can it be regarded as cognitively meaningful. So indeed, although Bohr never committed himself with respect to the concept of truth he must have conceived of it partly in these terms. For he held both that "in our description of nature the purpose is not to disclose the real essence of the phenomena but only to track down, so far as it is possible, relations between the manifold aspects of our experience" and also that the cognition of continuity characterizing our perception of phenomena constitutes the condition for the possession of unambiguous and objective knowledge of nature. These assumptions entail a notion of truth as coherence which is, of course, in no way in conflict with the verificationist notion of truth which underlies the positivistic theory of meaning, and both may be incorporated into an overall theory of meaning. How, in Bohr's view, this is to be achieved will be examined in the final chapter.

So the indefinability thesis which forms part of Bohr's notion of complementarity received at the end of the 1920s a purely epistemological justification quite consonant with the essential features of Høffding's philosophy. Atomic objects are real, Bohr says, but properties which we cannot observe without thereby exerting an influence on our observation of others which, with the former, jointly characterize an object exhaustively, exist merely in relation to specific observational circumstances. This is a consequence of the inseparability of the object and the measuring instrument. Of course, the inseparability factor in itself does not disprove a claim that these properties, in the absence of any intervention, exist in an undisturbed state. But Bohr has a further point: if empirical knowledge of one set of properties excludes every possibility of the acquisition of empirical knowledge of another, and if the mutually exclusive properties are related to the visualizability of an object in space and time and to its causal connections, to which a unified acquaintance is essential in order to ascribe objective existence to it, then no cognitive claim is made by saying that atomic objects have definite properties independent of the means by which they are measured. Thus, Bohr's conclusion is this: atomic objects have well-defined properties corresponding to classical state-defining parameters only to the extent that they are observed. It was this conclusion which Einstein regarded as a great challenge.

2. THE EPR DISCUSSION

In 1935 Bohr was confronted with the most serious of the attacks on his interpretation of quantum mechanics. One of the preconditions of this interpretation is that the quantum formalism is consistent – something which he had been able to prove in earlier discussions at the Solvay conferences in 1927 and 1930 with his most celebrated opponent, Einstein – but another is that it makes up a complete description of nature in the sense that there is no property of the atomic object which does not correspond to a parameter in the theory.[21] Bohr assumed that Heisenberg's uncertainty relations set the limits not only for the simultaneous measurability of dynamic and kinematic properties but also for the cognitive meaningfulness of a simultaneous ascription of these quantities to the object. This last, Bohr's indefinability thesis, was based on an epistemically grounded assumption according to which what is observed has to be causally connected in space and time in order that our experience be related to something objectively real. But since the trajectories of non-observed atomic objects cannot be described in terms of causal space-time modes of description there are no cognitive grounds for attributing dynamical and kinematical quantities to such objects independently of their observation. Thus, crucial to Bohr's argument for the indefinability thesis was his criterion of physical reality, which was entirely justified on epistemic grounds.

It was this criterion of reality which Einstein opposed, offering another which he thought was more in harmony with the aim of physics as well as being one which he believed could be supported by an ingenious physical argument. When Einstein, Podolsky and Rosen published their famous criticism of the completeness of quantum mechanics in a paper entitled "Can Quantum-Mechanical Description of Physical Reality Be Considered Complete?", they laid down two conditions, one stating when a physical theory can be characterized as complete, another stating when a physical quantity can be regarded as representing something real.[22] First they propose a necessary condition for the completeness of a theory:

(C) A physical theory is complete only if "every element of the physical reality has a counterpart in the physical theory".

Subsequently they suggest that a sufficient condition for a physical quantity to be real is

(R) "If, without in any way disturbing a system, we can predict with certainty (i.e. with probability equal to unity) the value of a physical quantity, then there exists an element of physical reality corresponding to this physical quantity".

That is, a physical theory can be considered as complete only if it can be demonstrated that every physical magnitude which can be assigned reality according to (R) has a representation in the theory. Thus, if it can be demonstrated that there exist physical quantities the values of which are predictable, but which are not accounted for by the theory, then one would have proved its incompleteness. Consequently, if quantum mechanics is complete, then complementary quantities cannot have simultaneous reality. But, in contrast, if it is possible to make a definite prediction of the values of two complementary quantities, such as the position and momentum of a particle, quantum mechanics cannot be complete. It was this last conditional which Einstein and his collaborators set themselves the task of proving.

With this as their aim they based their argument on the examination of a thought experiment involving the measuring and prediction of position and momentum of two particles A and B which had once interacted because, say, they had been created by a radioactive decay process, making the total momentum of the composite system zero. If, when A and B are sufficiently far enough apart for the measurement of one of them not to physically disturb the other, one would be able to measure A's momentum p, after which B's momentum $-p$ could be calculated using the conservation theorem of momentum. However, instead of using the measurement of A's momentum one might just as well have chosen to measure A's position q and then to have predicted B's position $-q$ on the basis of a theoretical definition of the state of the composite system. Since it is apparently open to one freely to choose between measuring A's momentum or its position, and then to calculate B's momentum or its position respectively, long after A and B have ceased interacting, there are fair reasons for believing that in both cases B would have possessed the same undisturbed physical state in the form of a definite position and a definite momentum. The previous interaction between A and B had simply established a correlation between their corresponding variables, and that the prediction of each of these conjugate parameters on the basis of knowledge of the corresponding parameter is possible ensures the satisfaction of condition (R). So both conjugate parameters are supplied with an element of reality. But since the simultaneous possession by complementary quantities of exact values is ruled out by quantum mechanics, this theory must be incomplete.

The EPR argument, as it is stated here, is far from compelling given the premisses. It is well-known that several further assumptions are required if the argument is to go through.[23] The most significant hiatus is where the simultaneous existence of B's position and momentum is inferred from the possibility of measuring either of these parameters with respect to A, since they cannot be measured simultaneously. Einstein and his collaborators could not merely rely on their criterion of reality to fill in the gap. They had to add an assumption which did not take for granted the simultaneous reality of conjugate parameters that they so desperately wanted to prove. In fact at the end of their paper they gave expression to their suspicion that the proof might be considered deficient in this respect.

> One could object to this conclusion on the grounds that our criterion of reality is not sufficiently restrictive. Indeed, one would not arrive at our conclusion if one insisted that two or more physical quantities can be regarded as simultaneous elements of reality *only when they can be simultaneously measured or predicted.* On this point of view, since either one or the other, but not both simultaneously, of the quantities of P and Q can be predicted, they are not simultaneously real. This makes the reality of P and Q depend upon the process of measurement carried out on the first system which does not disturb the second system in any way. No reasonable definition of reality could be expected to permit this.

But it was precisely such an "unreasonable" definition of reality which Bohr was going to defend.

In fairness it must be said that Einstein was dissatisfied with how the paper turned out. Don Howard has successfully substantiated this claim, showing that Einstein strengthened the original argument with an additional assumption to the effect that spatially separated systems possess their own independent real states.[24] However, Bohr did not know of Einstein's reservation nor his more sophisticated argument at the time he prepared his reply to the EPR paper, partly because Einstein did not make an explicit distinction between separability and locality, and conflated both concepts in his 'separation principle'[25] until long after he had studied Bohr's reply to the EPR paper. There is no later evidence showing that Bohr actually saw any differences between the original EPR argument and Einstein's own. Quite the contrary, in 1949 when he wrote about his discussion with Einstein he merely repeated the arguments which were generated by his earliest reaction to the EPR paper. But even if

Bohr were to have acknowledged some differences there is no reason to believe that it would have had a substantial effect on his argument.

The Einstein, Podolsky and Rosen paper seems to have shaken Bohr for a while. A rather dramatic account of his reaction to their paper has been given by Rosenfeld, who worked closely together with him at that time.

This onslaught came down upon us as a bolt from the blue. Its effect on Bohr was remarkable. We were then in the midst of groping attempts at exploring the implications of the fluctuations of charge and current distributions, which presented us with riddles of a kind we had not met in electrodynamics. A new worry could not come at a less propitious time. Yet, as soon as Bohr heard my report of Einstein's argument, everything else was abandoned: we had to clear up such a misunderstanding at once. We should reply by taking up the same example and showing the right way to speak about it. In great excitement, Bohr immediately started dictating to me the outline of such a reply. Very soon, however, he became hesitant: "No, this won't do, we must try all over again ... we must make it quite clear ...". So it went on for a while, with growing wonder at the unexpected subtlety of the argument. Now and then, he would turn to me: "What *can* they mean? Do *you* understand it?" There would follow some inconclusive exegesis. Clearly, we were farther from the mark than we first thought. Eventually, he broke off with the familiar remark that he "must sleep on it". The next morning he at once took up the dictation again, and I was struck by a change in the tone of the sentences: there was no trace in them of the previous day's sharp expressions of dissent. As I pointed out to him that he seemed to take a milder view of the case, he smiled: "That is a sign", he said, "that we are beginning to understand the problem". And indeed, the real work now began in earnest: day after day, week after week, the whole argument was patiently scrutinized with the help of simpler and more transparent examples. Einstein's problem was reshaped and its solution reformulated with such precision and clarity that the weakness in the critic's [sic] reasoning became evident, and their whole argumentation, for all its false brilliance, fell to pieces. "They do it smartly", Bohr commented, "but what counts is to do it right".[26]

I have quoted the entire passage because it illustrates quite well, without that perhaps being the author's intention, the difficulties which Bohr felt that his philosophy had been plunged into by the EPR argument.

Thus, as soon as he heard about the paper Bohr put all other work aside in order to work out an answer. After two months he published his first response in the form of a short "letter to the editor" in *Nature*, entitled "Quantum Mechanics and Physical Reality",[27] and after three more months he published in *Physical Review* a paper with the same title as Einstein, Podolsky and Rosen's: "Can Quantum-Mechanical Description of Physical Reality be Considered Complete?".[28] As one would have expected from an uncritical admirer of Bohr, Rosenfeld adds the following to his vivid but also merciless description of Bohr's preparation of a counterblast: "The refutation of Einstein's criticism does not add any new element to the conception of complementarity". In my opinion, however, Bohr intuitively grew to realize, during the period of which he was working on his reply, that his epistemic defence of the indefinability thesis was inadequate. If this is true it explains why it took him two months to finish it,[29] and why after a good night's sleep Bohr calmed down, changing "the tone of the sentences" while he took "a milder view of the case".

The core of Bohr's answer was, indeed, a criticism of the criterion of reality (R) suggested by Einstein and his co-workers, as can be seen from the following passage:

> The apparent contradiction in fact discloses only an essential inadequacy of the customary viewpoint of natural philosophy for a [causal/rational] account of physical phenomena of the type with which we are concerned in quantum mechanics. Indeed the *finite interaction between object and measuring agencies* conditioned by the very existence of the quantum of action entails – because of the impossibility of controlling the reaction of the object on the measuring instruments, if these are to serve their purpose – the necessity of a final renunciation of the classical ideal of causality and a radical revision of our attitude towards the problem of physical reality. In fact, as we shall see, a criterion of reality like that proposed by the named authors contains – however cautious its formulation may appear – an essential ambiguity when it is applied on the actual problems with which we are here concerned.[30]

This passage is also reproduced by Bohr in his "Discussion with Einstein on Epistemological Problems in Atomic Physics" in a slightly different version. The main difference, which is fundamentally no difference at all, is that the term "causal" in the first sentence has been replaced by the term "rational". There is no better way of indicating that Bohr regarded the causal account of the physical phenomena as equivalent to the rational account. Such an account constitutes the criterion of reality, Bohr would say, but since its application in quantum mechanics is severely restricted because of "the necessity of a final renunciation of the classical ideal of causality", it follows that we are forced to accept "a radical revision of our attitude towards the problem of physical reality".

In his response to the EPR argument Bohr mentions the fact that the finite interaction between the object and the measuring instrument has to be taken into consideration. But, no matter what, the EPR experiment seems to present us with a theoretical situation in relation to which it is unjustifiable to talk about the uncontrollable interaction between object and instrument. Bohr cannot, it seems, appeal to the disturbance of the object A by the measuring instrument as an explanation of why a causal spatio-temporal description of the unmeasured object B cannot be sustained, simply because it is assumed by the authors that the measurement of the value of one of A's conjugate parameters cannot affect the value of B's corresponding parameters. A and B do not physically interact, hence the disturbance of A's states by the process of measurement cannot physically influence B's states. If this is so, how can Bohr make a point of declaring that the criterion of reality proposed by Einstein and his collaborators contains an ambiguity with respect to the meaning of an expression such as "without in any way disturbing a system" as long as his own criterion of reality seems to be vulnerable to the implications of the EPR thought experiment?

Bohr concedes, of course, that mechanical disturbance of the system is not involved when we calculate, for instance, B's momentum after observing A's. And yet, he says, the choice of measuring A's momentum has an influence on the conditions which define the possible type of predictions of the future behavior of the system.

Of course there is in a case like that just considered no question of a mechanical disturbance of the system under investigation during the last critical stage of the measuring procedure. But even at this stage there is essentially the question of *an influence on the very conditions which define the possible types of predictions regarding the future behavior of the system.* Since these conditions constitute an inherent element of the description of any phenomenon to which the term "physical reality" can properly be attached, we see that the argumentation of the mentioned authors does not justify their conclusion that quantum-mechanical description is essentially incomplete.[31]

Indeed, by saying "an influence on the very conditions" Bohr does not mean a physical influence. In fact, the EPR argument weakens the force of, if it does not deprive Bohr of, one of the main premisses of the epistemic argument for complementarity: that the measuring instrument disturbs the behavior of the object. He shares with Einstein the assumption that all causal actions are local and that it takes time for a physical signal to move from one place to another. Bohr has to admit that the choice of measuring either A's momentum or position does not have any physical influence on which property B possesses, and the correlation between one of A's conjugate parameters with the corresponding one of B's involves no physical disturbance. Both he and Einstein agree that A and B have space-like separation. But what he denies is that two complementary wave functions, which are eigen-functions of the position and the momentum operator with the values q and p, respectively, describe *one and the same reality*, which is what Einstein holds. As we shall see in the following section, this concession forces Bohr to change some of his terminology as well as part of the underlying argument for the indefinability thesis. The emphasis which Bohr now puts on the "very conditions which define the types of possible predictions" introduces a new form of argument which seems to be Bohr's own, and which reflects his reaction to the EPR argument.

What Bohr now wants to argue is that the cognitively meaningful application of a concept presupposes the fulfillment of certain experimental conditions, since their satisfaction is the only criterion by which it can be decided whether or not the concept is applied correctly. This also holds in the case of the EPR experiment. That a choice may be made between measuring A's position or its momentum does not change the situation radically with respect to the ascription of momentum or position to B because it makes sense to ascribe these properties to B if, and only if, one is actually capable of attributing such properties to A; and one is capable of this only if the conditions of a position measurement or of a momentum measurement are realized.

In an attempt to strengthen this conclusion Bohr simulated the EPR experiment by imagining two particles with a certain initial momentum each passing through its own slit in a diaphragm. On the basis of an accurate measurement of the momentum of the diaphragm before and after the passage of the particles the sum of each particle's momentum perpendicular to the slits may be ascertained, as may too the difference of their positional coordinates in the same direction from the distance between the two narrow slits. Bohr's point is then that one cannot acquire knowledge of both position and momentum of either of the two particles. What is possible is to get to know either the position or the momentum of each of the two particles since subsequent to their passage we have to make a choice between measuring either the position of one of them or its momentum, and then we are able to calculate the corresponding parameter on the basis of the knowledge we have acquired of the compound system.

It might be thought that there is one way in which we might repudiate Bohr's criterion. It might be suggested that if a measurement determines A's momentum on the basis of which B's momentum may be calculated, it must be possible at exactly the same time to determine B's position and then on the basis of that calculate A's position. This means that even if Bohr is right in his philosophical assumption that the application of concepts such as position and momentum in quantum mechanics is meaningful only if certain empirical conditions are realized, he is nevertheless forced to concede that it is possible to construct a situation in which we have sufficient empirical and theoretical warrants for ascribing a position and a momentum to a particle. That is, even if Bohr's conception of reality in terms of what is empirically assertable is accepted, it cannot be denied that, by having had their corresponding parameters correlated through the previous interaction, A and B must both have an exact position and an exact momentum. This is because Bohr indeed accepts both that it is possible to calculate B's momentum theoretically on the basis of the conservation theorem whenever A's momentum is determined by observation and that it is possible to calculate B's position theoretically whenever A's position is measured. But Cliff Hooker has demonstrated that such an objection is impotent.[32] In an attempt to measure, say, A's position while simultaneously attempting to measure B's momentum an unknown amount of momentum is transferred from A to the common laboratory frame of reference and thus interferes with the momentum measurement of B.

Thus Bohr believes that in quantum mechanics we are only epistemically justified in ascribing a kinematical or a dynamical property to what can be directly observed. He has, therefore, to dismiss Einstein's criterion of reality, which entails that a particle possesses an exact position and an exact momentum in spite of the fact that it is only the value of one of these quantities which can be determined experimentally at a time. Bohr rejects it because he thinks that in quantum mechanics the experimental arrangement constitutes a necessary condition for the meaningful ascription of these attributes. It makes little sense to talk about a well-defined state, as Einstein does, unless the conditions which are necessary for the ascription of the properties which purportedly define such a state are satisfied. The conditions are not satisfied in the situation where Einstein and his collaborators assume that a particle, in case B, has one and only one physical state by simultaneously ascribing it an exact momentum and exact position. The necessary conditions for the application of the concept of momentum or the concept of position are fulfilled whenever the experimental conditions for measuring them are realized, and since the conditions are mutually exclusive, it follows that we have no epistemic grounds for ascribing momentum and position to one single state. The reason, according to Bohr, we find ourselves in quantum mechanics in a situation in which a state is well-defined only if there exists an empirical warrant for the application of the proper concepts is because the criterion of reality, that is, the existence of a causal space-time description which normally guarantees that it is cognitively meaningful to talk about well-defined, unobserved states, finds no application in the domain of quantum mechanics. Bohr's reply in *Physical Review* was more philosophical than physical and neither here nor elsewhere did he really explain how the physics of the composite system should be understood. But since Bohr emphasized that the quantum of action constitutes an unanalyzable whole one might expect that he would also say something similar about the composite system of A and B. What this involves can best be seen in the light of the debate which has followed in the wake of Aspect's and others experiments on testing the Bell inequality.

In 1964 J.S. Bell succeeded in showing that it was possible to distinguish empirically between models containing so-called hidden variables, which are thought of as making quantum mechanics a complete description of nature, and models presupposing orthodox quantum mechanics. Bell proved that all deterministic hidden variable models which can be shown to satisfy a certain reasonable locality condition would respect a certain theorem which statistical predictions by quantum mechanics violate. The condition employed bears upon the assumption that there exists no action-at-a-distance influence. Later Bell generalized his theorem to stochastic hidden variable models by introducing a more general locality condition. Since then Jon Jarrett has argued that this stronger locality condition in fact consists of two logically independent conditions.[33] One of the conditions, which may be called the locality principle, is a condition to the effect that no signal with a velocity greater than light can be a part of any correlations between A and B. That is to say, the measurement of states of A and of B are physically independent of each other. The other condition, which might be called the completeness assumption, states that the probability of an observable, say, a of A to possess a certain value is dependent on a hidden variable λ but independent of the value possessed by an observable b of B. However, it can be shown that the latter condition is equivalent to the separability principle in deterministic hidden variable theories: namely, that spatially separated particles, like A and B, possess individual physical states whose properties are definite and well-defined.[34]

Bell's work can be seen as a development of the EPR argument, although he did not base his reflections on the EPR experiment but on a spin experiment which was originally suggested by Bohm and Aharonov. In this experiment the position and momentum of a pair of twin particles have been replaced by two spin components (say x and z) of a pair of spin-(1/2) particles which are emitted on a singlet state with a total spin of zero, but which thereafter have become separated by a space-like interval. One may either measure the x-component of A s_x = -(1/2), and hence ascribe a wave function to B which is an eigenfunction of the s_x-operator with the value of s_x = +(1/2), or measure the z-component of A s_z and thus ascribe another wave function to B, which is now an eigen-function of the s_z-operator. Indeed, Bohr's interpretation of orthodox quantum mechanics is that these two wave functions are complementary so they do not apply simultaneously to the same system. Nevertheless, Bell could prove that there would be a difference in the statistical outcome of this experiment according to whether the prediction was made within models of local hidden variables or models of orthodox quantum mechanics. The difference finds its expression in an inequality which indicates a certain statistical correlation between many measurements of such pairs of twins. Finally, in 1982, this inequality was tested most convincingly by Alain Aspect and collaborators, who used pairs of polarized photons to discover whether or not the outcome was in agreement with orthodox quantum mechanics, and they obtained a result which was in agreement with the predictions of this theory. Consequently, theories containing local hidden variables seem to have been falsified.

Notice, first of all, that the interpretation of the outcome of various experimental tests of Bell's theorem may involve only one of the two questions which has haunted the philosophy of quantum mechanics from the very beginning. For we may distinguish between the question concerning the causal correlation of spatially separated states, which we may call the problem of locality, and the question regarding the conditions under which the ascription of dynamical properties to an atomic object is meaningful, which may be labelled the problem of measurement. It is these two problems which stand as the most serious challenges to any realistic understanding of quantum mechanics, and it may be claimed that they are logically distinct. So a philosophical solution to one of them need not go very far with the other. But such a view would be opposite to Bohr's. If we look upon the two questions as distinct the results of Aspect's experiments seem to force us to give up the principle of locality, but if we, as Bohr did, regard them as being intimately connected one has to abandon the principle of separability.

Whether it is the locality principle or the separability principle which has to be sacrificed cannot be determined by philosophical discussion. As we have seen, Bohr apparently held a position to the effect that the compound microsystem consisting of A and B which is described by one and the same wave function constitutes an indivisible unity. The existence of such an indivisible whole excludes every assumption that the composite system is in a definite but unknown state. That is, the state vector represents in terms of physics the entire system as an objective but undetermined superposition which can only be dissolved if it is reduced to one of its ground states through a position or momentum measurement. This is possible only because canonically conjugate parameters do not, according to Bohr, stand for any intrinsic state properties but for objective relational ones. Consequently, if he is right, it is the separability principle which has to be abandoned.

This conclusion falls in line with his view that there are similarly nonseparable states resulting from the interaction between the atomic phenomena and the measuring instrument. Just as the state of the object and the state of the instrument are dynamically inseparable, so too the individual states of a pair of coupled particles cannot be considered in isolation. As Bohr said a few years later when alluding to the EPR argument in his Warsaw lecture:

In fact, the paradox finds its complete solution within the frame of the quantum mechanical formalism, according to which no well-defined use of the concept of "state" can be made as referring to the object separate from the body with which it has been in contact, until the external conditions involved in the definition of this concept are unambiguously fixed by a further suitable control of the auxiliary body.[35]

Thus, in the EPR experiment one of the two objects, the auxiliary body A, may be treated as an instrument and the other, B, treated as an object, and the states of a pair of objects which have once interacted are related to each other in the same way as the state of an object and the state of an instrument are during the process of measurement. However, this correspondence is not due to a coincidental causal connection. For in that case it ought to be possible to give a definition of a state of B that is independent of a definition of the state of A. It is senseless to talk about a causal connection between either a pair of EPR objects or an atomic object and the measuring instrument if the concept of a state cannot be applied to an object independently of the application of this concept on the auxiliary object or the instrument. This is probably what Bohr has in mind. That is, it is impossible to define a state of an object B unless another has been defined of another object A, and this is possible when and only when the experimental conditions for ascribing to A a certain state are specified. It also means that from the knowledge of the value of the state of A, being an atomic object or an instrument, we may infer the value of the state of the object B.

But in the case of the abandonment of the principle of separability, owing to the rejection of the intrinsic property theory, it is difficult to see what it is that might really justify the individuation of A and B. Quantum mechanics makes the assumption that there is a pair of objects which form one single indivisible system, not that there are two objects each with their own distinct states. For once A and B have interacted with each other the system they jointly compose has to be described by a state vector which does not contain a product of the state vectors of its components. The questions concerning individuation are therefore these: Do we create two individual objects at the moment at which we determine their states by measuring the value of a conjugate parameter in one spatial area of the composite system? Or do we produce two simultaneous individual states of one and the same system consisting of two non-separated objects? Or do we in fact establish the conditions for the occurrence of two distinct states of a pair of objects which have existed as separate individuals all along? Or do we establish the experimental context in which we can talk about individual atomic objects? And what argument do we have for asserting the truth of one of these alternatives? What sense does it make to talk about *one* system consisting of two spatially separated objects with non-separable states? And what does it mean to say of two objects that they are numerically different but yet do not possess well-defined states of their own until one of them has been measured? Bohr would clearly say none of this but that it is illegitimate to assume that atomic objects have a separate identity independent of a particular context in which the experimental arrangement is fully specified and in which the states of the objects are made evident. Thus, the notion of a criterion for identity over time remains a problematic one for Bohr so long as he denies the existence of intrinsic state-properties.

It may be objected, however, that what he denies is that classical concepts can refer to free isolated atomic objects, not that these do not possess intrinsic state-properties. But such a claim is wrong. As I shall argue in the final chapter, there is no evidence which supports the claim that Bohr should have thought of atomic objects as possessing non-classical, intrinsic state-properties – quite the reverse. If this is so, it is impossible to see how atomic objects, following Bohr's view, can be individuated in situations where the classical concepts do not apply.

Instead of tampering with the principle of separation in order to bring it into harmony with an intelligible principle of individuation, one might therefore be tempted to give up the locality principle, arguing that it is perfectly acceptable to suggest that it be violated. In this case the creation of a physical theory is required which can explain the outcome of Aspect's experiment and similar kinds of phenomena. Personally, I am much in favor of this latter alternative, which I believe to be both conceptually and physically sound.[36]

However, the philosophical disagreement between Bohr and Einstein about the conditions for the description of atomic objects is totally unaffected by the question concerning which of these principles is the right one. Since both took the locality principle for granted, it was only natural that their philosophical disagreement implied divergent opinions with regard to the separability principle, since Einstein thought of atomic objects as having intrinsic properties, and hence individual states, whereas Bohr denied the existence of such independent properties or states. But both might, logically, also have given up the locality principle instead of the separability principle while still holding different views on what constitutes a condition for the description of nature. Bell's work and the result of Aspect's experiments are logically irrelevant to the part of Bohr's answer which concerns his theory of meaning, although historically the EPR argument had a strong impact on Bohr's later expression of complementarity and underlying arguments. The fundamental difference between these two giants was not so much one concerning physics as it was one concerning the criterion for the possession by statements purportedly about reality of a cognitive content.

3. THE CONDITIONS FOR DESCRIPTION

The EPR argument was a reaction to the following philosophical query: if the atomic object is disturbed, as Bohr claimed in the late twenties and early thirties, by the measuring instrument whenever we are observing it, then it is quite possible that the object has both a definite momentum and a definite position before and after the act of measurement, though we are not, in principle, in a position to ascertain anything empirically about the value of one of the conjugate variables if we have exact knowledge of the value of the other. It might be conceded that it is not simultaneously ascertainable that both these properties obtain and yet still be asserted that atomic objects always possess definite properties. Nothing of what Bohr had said up to 1935 about the implications of the quantum of action for the results of observation precluded anyone from adopting the same attitude with respect to the feasibility of the acquisition of empirical knowledge, while combining this view with a realistic ontology: that is, by holding that objective truth may transcend our capacity for making empirical judgments. On the basis of such an approach it might be hoped that it would some day be possible to become simultaneously acquainted with the two conjugate parameters on a purely theoretical basis, regarding Heisenberg's uncertainty relations as mere formal statements that express certain epistemological constraints by indicating the limits with respect to the acquisition of observational knowledge of kinematic and dynamic variables in the domain of atomic objects.

It was not until Einstein, Podolsky and Rosen challenged the completeness of quantum mechanics that Bohr was forced to take such an approach seriously, and gradually he began to recognize that his previously held epistemic argument in support of complementarity might be inadequate to meet the arguments constituting the challenge of such a position. From then on he attempted to strengthen the arguments for his own point of view by appealing more and more to what I shall call the semantic argument for the indefinability thesis: an argument which has the same epistemological and ontological implications as the epistemic argument has but which is not vulnerable to the objections of Einstein, Podolsky and Rosen. Thus it is not enough for Bohr to argue that since we cannot measure a pair of canonically conjugate parameters simultaneously we have no cognitive grounds for holding that an atomic object possesses well-defined properties. This claim carries no weight given the criterion of reality endorsed by his opponents. In fact he had to turn the argument upside down: which is to say that since we cannot simultaneously define exact values of conjugate variables (complementary concepts not being at one and the same time meaningfully ascribable to atomic objects), we cannot measure these exact values simultaneously. In his reply to Einstein, Podolsky and Rosen, as we have seen, Bohr still expresses himself ambiguously, while in his contribution to *Albert Einstein: Philosopher-Scientist* and in other papers from the forties and fifties he develops the semantic argument in as much as he now considers a reference to the entire experimental arrangement as being what determines the conditions for the correct use of complementary concepts. Or to rephrase it in a modern jargon, Bohr now argues that a reference to the entire experimental set-up enters into the specification of the truth conditions for any statement involving the Heisenberg indeterminacy relations, which express the scope of the ascription of an exact momentum and an exact position to a quantum mechanical system. However, this claim does in fact imply that it is incoherent to talk about the disturbance of the object by the measuring instrument as such but still allows locutions to the effect that the integral interaction between the atomic object and the measuring instrument sets limits to a simultaneous ascription of momentum and position.

So the debate between Einstein and Bohr had a direct impact on the respective position of the two combatants. Eventually Einstein recognized that one might dismiss his criterion of reality if one denied, as Bohr did, the principle of separability. In fact Einstein once wrote, "Of the 'orthodox' quantum theoreticians whose position I know, Niels Bohr's seems to me to come nearest to doing justice to the problem".[37] And little by little Bohr made certain fundamental revisions in his terminology as well as with respect to the underlying philosophical arguments of complementarity in order to take the sting out of the objections contained in the EPR paper. Most commentators have perceived no significant alteration in Bohr's philosophy of physics at all, the changes wrought in expression and emphasis notwithstanding.[38] The structure of the exposition of Bohr's thought by these philosophers may be characterized as being more synchronous than diachronous in form, in the sense that quotations from the earliest essays occur alongside ones from the latest essays. Such an approach blurs to a certain extent the question of whether or not differences in form of expression in fact signal substantive changes in the foundations of his philosophy. Only a few commentators have seen a really radical change in Bohr's philosophy.[39]

I shall argue that Bohr did not modify his philosophy dramatically and that what he altered were some of the basic arguments in support of it. After 1935 his grounds for asserting complementarity were not so much epistemological as they were conceptual or semantical. And I see two reasons for this alteration. First, as I have just argued, I think that the EPR thought experiment posed a serious challenge to Bohr's philosophy which he could not tackle by appealing solely to the epistemic argument for the indefinability thesis. Second, Høffding's direct influence had come to an end with his death a few years earlier, with the result that Bohr emerged as a philosopher in his own right.

Now Bohr wanted to argue that the formalism of quantum mechanics reflects the fact that atomic objects cannot, for instance, simultaneously be attributed a definite position and a definite momentum like objects in classical mechanics because these concepts are not well-defined under the same circumstances. Similar constraints hold for other conjugate variables. The use of concepts which enter into the description of atomic objects requires the satisfaction of certain conditions in order for the concepts to be well-defined. Of course in classical mechanics as well the meaningful application of descriptive concepts requires the realization of certain conditions. But the conditions for description in quantum mechanics differ from those in classical mechanics even though the concepts which are used for descriptions are the same in both cases, because the latter conditions are not universally obtainable as it was once believed. An analysis of the conditions shows that only commuting observables can simultaneously be well-defined and have an exact value. If observables, on the other hand, are specified by two Hermitian operators which do not commute, then the process of measurement cannot yield a definite value of both because the conditions for assigning concepts designated by these operators exclude each other. This stands in sharp conflict with classical mechanics, which does not set any limit to the precision with which we may simultaneously assign a value to the corresponding quantity. According to classical mechanics a particle may be attributed a definite momentum and a definite position no matter whether these are measured or not since the conditions for a precise definition of one of these quantities are not incompatible with those for a precise definition of the other. But this does not hold true in quantum mechanics, where position and momentum as well as time and energy are non-commuting observables: they cannot be defined simultaneously owing to the fact that the conditions for a correct application of such observables cannot be realized at the same time.

In "Causality and Complementarity" from 1937, the first philosophical paper written by Bohr subsequent to his reply to the EPR paper, he warns against the misunderstanding one is a victim of if one believes that Heisenberg indeterminacy relations merely set limits to the accuracy obtainable by simultaneous measurement of the position and momentum of a particle.

> According to such a formulation it would appear as though we had to do with some arbitrary renunciation of the measurement of either the one or the other of the two well defined attributes of the object, which would not preclude the possibility of a future theory taking both attributes into account on the lines for the classical physics. From the above considerations it should be clear that the whole situation in atomic physics deprives of all meaning such inherent attributes as the idealizations of classical physics would ascribe to the object.[40]

The indeterminacy relations are formal expressions of the logical possibility of applying classical concepts such as position and momentum to the domain of the quantum of action if they are to be used in a well-defined manner, and if the scope of their application is to be determined by the choice of the experimental arrangement.

Again in the Warsaw paper from 1938, Bohr acknowledged quite explicitly the inadequacy of the epistemic argument for the indefinability thesis. The reason why it is impossible to ascribe exact values to each of two conjugate but non-commuting variables is not because it is impossible to measure them both at the same time. Indeed it is. But,

> the statistical character of the uncertainty relations in no way originates from any failure of measurements to discriminate within a certain latitude between classically describable states of the object, but rather expresses an essential limitation of the applicability of classical ideas to the analysis of quantum phenomena.[41]

In his reply to the EPR paper Bohr had been more vague about whether it is ignorance of the exact value of one of a pair of conjugate parameters which is a sufficient (and necessary) ground for claiming the indefinability of this quantity, or whether it is the impossibility of a simultaneous and yet significant ascription of complementary concepts to an atomic object which is a sufficient (and necessary) ground for the simultaneous unmeasurability of complementary quantities. He says,

Indeed we have in each experimental arrangement suited for the study of proper quantum phenomena not merely to do with an ignorance of the value of certain physical quantities, but with the impossibility of defining these quantities in an unambiguous way.[42]

Nonetheless, I take the epistemic argument to be that to which he may still be appealing. But shortly afterward Bohr grasped the full significance of the semantic argument. This argument for the indefinability thesis is clearly expressed by Bohr in a passage to be found in an essay from 1958, published as "Quantum Physics and Philosophy", where he discusses the renunciation of the simultaneous determination of kinematic and dynamic variables which is required for the definition of the state of a system and which is expressed by Heisenberg's indeterminacy relations.

In fact, the limited commutability of the symbols by which such variables are represented in the quantal formalism corresponds to the mutual exclusion of the experimental arrangements required for their unambiguous definition. In this context, we are of course not concerned with a restriction as to the accuracy of measurements, but with a limitation of the well-defined application of space-time concepts and dynamical conservation laws, entailed by the necessary distinction between measuring instrument and atomic objects.[43]

That two conjugate variables are not simultaneously measurable follows from the indefinability thesis, not the other way around, and it in turn is due to the non-satisfaction of the conditions for a simultaneous application of the concepts in question, which again follows from the distinction between the atomic object and the measuring instrument.

Quite symptomatically a change in the underlying argument is signalled by certain alterations in Bohrs terminology. In the Como paper as well as in the other papers from before 1935 Bohr spoke of "Heisenberg uncertainty relations" and of the uncertainty of our knowledge of either position or momentum, thereby indicating that these relations primarily express what can be known and only secondarily what can be defined.[44] Given the uncertainty of knowledge, the indeterminacy of concepts follows, so to speak. This reading may find some support in the *Introductory Survey*, where Bohr first mentions that Heisenberg had pointed out the close connection between the limited applicability of mechanical concepts and the element of uncertainty which figures in our knowledge of the course of the phenomenon, owing to the interacting measuring instrument, and then adds that *this* indeterminacy prevents the simultaneous use of space-time concepts and the laws of conservation of energy and momentum.[45] But after his confrontation with the EPR paper Bohr began to put the matter differently by calling the formal relations "Heisenberg indeterminacy relations",[46] thereby stressing that the question concerning the applicability of concepts is logically prior to that concerning the measurability of the conjugate variables which these concepts denote. Even in the Danish versions of his papers we find a similar change of terminology around 1935 between "*usikkerhedsrelationer*" and "*ubestemthedsrelationer*" (where the Danish terms are equivalent to the respective English terms), suggesting a real change in Bohr's thinking and not in how he expressed his point in English. He continued subsequently to hold that it is the indeterminacy of concepts that entails the uncertainty of knowledge.

To see how, let us now look more closely at the structure of the semantic argument. In the most important epistemological work written by Bohr at this stage of his life, the "Discussion with Einstein on Epistemological Problems in Atomic Physics" from 1949, and in other essays, he stresses several times the necessity of maintaining "the distinction between the *objects* under investigation and the *measuring instruments* which serves to define, in classical terms, the conditions under which the phenomena appear".[47] Recall that this distinction derives from the fact that "*however far the phenomena transcend the scope of classical physical explanation, the account of all evidence must be expressed in classical terms*".[48] But since quantum objects are ruled by the quantum of action it is only the description of experiments and the evidence thus provided by them which satisfy the conditions for unambiguous communication. As Bohr remarks in the continuation:

> The argument is simply that by the word "experiment" we refer to a situation where we can tell others what we have done and what we have learned and that, therefore, the account of the experimental arrangement and of the results of the observations must be expressed in unambiguous language with suitable application of the terminology of classical physics.[49]

To put it another way: in spite of a refinement of terminology in quantum mechanics in relation to the classical vocabulary, based on empirical and theoretical considerations, all unambiguous communication about our sensory experience is "ultimately based on common language, adapted to orientation in our surroundings and to tracing relationship between cause and effect".[50] This means that the conditions for unambiguous communication of any common sensory experience are constituted by the presence of causal connections between phenomena in space and time. Indeed a causal spatio-temporal description of this experience is a precondition for experience to be concerned with what is independent of the communicating subject, and that type of description is necessary for coherent statements and unambiguous communication. "The description of ordinary experience presupposes the unrestricted divisibility of the course of the phenomena in space and time and the linking of all steps in an unbroken chain in terms of cause and effect".[51] Thus the very existence of a language for unambiguous communication presupposes that there is a clearly defined object with which communication is concerned, but such a condition is satisfied only if the object of communication is separable in a well-defined manner from the subject who communicates.[52] So the causal spatio-temporal description of the measuring instrument is what gives us the distinction between the experimental set-up as an object and the subject who describes it. However, if an object is to be characterized as a measuring instrument it demands a further distinction between what is measured and what does the measuring. Thus any unambiguous communication of a state of an object being measured by a measuring instrument requires a distinction between the object which is in the relevant state and the object which records this state. But if such a separation cannot be carried out, as it cannot in quantum mechanics where the description of atomic objects does not satisfy the classical conditions for unambiguous communication, it follows that in order to gain an unambiguous description of the evidence, the conditions necessary for a coherent application of the concepts of position and momentum must be those parts of sensory experience which show us that the simultaneous application of these concepts is incompatible, which conditions are constituted by the experimental arrangement.

The argument is evidently to the effect that the measuring instruments constitute the conditions for any unambiguous use of the quantum-mechanical formalism, whereby they serve to determine the conditions under which the phenomena meaningfully may be said to occur. One might say that a specification of the experimental arrangement is what constitutes the truth conditions of statements containing an ascription of a definite position or a definite momentum to an atomic object.

> Thus, a sentence like "we cannot know both the momentum and the position of an atomic object" raises at once questions as to the physical reality of two such attributes of the object, which can be answered only by referring to the conditions for the unambiguous use of space-time concepts, on the one hand, and dynamical conservation laws, on the other hand.[53]

The experimental conditions which determine the correct use of complementary concepts such as momentum and position are what the truth conditions are about.

That this is the case emerges in the following, where Bohr refers to the foregoing discussion about the consistency of the quantum-mechanical formalism with respect to the non-simultaneous measurement of momentum and position, stated by Heisenberg's indeterminacy principle, and displayed by Bohr's well-known thought experiments:

> Incidentally, we may remark that, for the illustration of the preceding considerations, it is not relevant that experiments involving an accurate control of the momentum or energy transfer from atomic particles to heavy bodies like diaphragms and shutters would be very difficult to perform, if practicable at all. It is only decisive that, in contrast to the proper measuring instruments, these bodies together with the particles would in such a case constitute the system to which the quantum-mechanical formalism has to be applied. As regards the specification of the conditions for any well-defined application of the formalism, it is moreover essential that the *whole experimental arrangement* be taken into account.[54]

Bohr seems here to add a further argument in his attempt to show that the application of the quantum mechanical formalism, and thereby the use of classical concepts in the domain of the quantum of action, is determined by certain appropriate experimental conditions if it is to be meaningful. Thus, in opposition to the notion of the uncontrollable exchange of energy and momentum between object and instrument to which he subscribed earlier and which provided him with an alibi for talking about a disturbance, Bohr now states that such a notion is irrelevant or rather, as I would say, meaningless because the application of the quantum mechanical formalism is not always confined to the microphysical system. Instead it is sometimes a question of certain parts of the measuring instrument interacting with the microphysical object in such a way that these parts are to be reckoned, together with the object, as part of the quantum mechanical system itself. Bohr underlined this very clearly when he wrote:

> The discussion ... thus emphasized once more the necessity of distinguishing, in study of atomic phenomena, between the proper measuring instruments which serves to define the reference frame and those parts which are to be regarded as objects under investigation and in the account of which quantum effects cannot be disregarded.[55]

When we are thus unable to distinguish between the atomic object itself and the parts of the experimental set-up, but must include both in the quantum mechanical system, the conditions under which the quantum mechanical formalism applies meaningfully must be related to the entire experimental situation and not merely to some putative features of the atomic object itself. The following and other statements by Bohr should be viewed in the light of this argument: "*... the impossibility of any sharp separation between the behavior of atomic objects and the interaction with the measuring instruments which serve to define the conditions under which the phenomena appear*".[56] So the movable partition between the quantum mechanical system and the measuring instrument also says something about our inability to make a logical distinction between the phenomenon considered as the effect of the atomic object's interaction with the measuring apparatus and the conditions under which it appears, which are the observational conditions that determine the correct use of classical concepts.

The central idea which Bohr developed in the late thirties, forties and fifties was therefore the idea that ordinary language, supplemented by the technical vocabulary of physics, serves the purpose of describing the world around us, and that through an analysis of the experiences described in that language we acquire a grasp of the most general conditions that make description of the world possible. The condition for the description of the ordinary experience of inanimate nature in terms of classical mechanics consists in the fact that any state of the system may be subdivided an indefinite number of times into increasingly narrowly defined states, thereby allowing the separability of the states of the object from the states of the measuring instrument. This is the condition for the well-defined application of causal and spatio-temporal concepts to ordinary experience and is the justification of our being able to speak about this experience in a rational and unambiguous way. However, the description of experience in contexts involving quantum mechanics does not satisfy this classical condition. The condition for unambiguous communication of such experience has to be replaced by other conditions in which the concepts space, time and causation can still be used in a well-defined manner. And the only conditions which satisfy this requirement are such observational situations where concepts such as position and momentum do not apply simultaneously, stemming from the fact that an experimental arrangement which allows a well-defined application of one of the concepts excludes another experimental arrangement in which the other concept would be well-defined. Therefore the experimental arrangement itself constitutes the conditions for the description of the atomic object; that is, the experimental arrangement determines the conditions for our linguistic behavior in domains where the quantum of action plays a role, in the sense that any successful reference to it guarantees that the use of concepts such as position and momentum be unambiguous, even in this context of experience.

From Bohr's point of view what follows is the conclusion that any proper understanding of the quantum mechanical formalism necessarily involves reference to the experimental set-up, since it is the observational conditions necessary for the appearance of the phenomenon which possess the properties to which the observational terms in all quantum mechanical predictions refer. Furthermore, since it is impossible to separate the behavior of the atomic object from its interactions with the measuring device, we cannot distinguish between the object itself and parts of the experimental set-up.

Hence it makes no sense, Bohr says, either to speak of "disturbing the phenomena by observation" or "the creation of the physical attributes of objects through measurements", because such expressions would hardly harmonize with "common language and practical definition".[57] The emphasis of the meaninglessness of a disturbance language, however, runs contrary to what he himself had said earlier, and this change of attitude can be seen as an immediate reaction to the EPR paradox. But although Bohr after 1935 regarded any talk of the disturbance of the object by measurement as senseless, by then understanding this expression as referring to some mechanical influence between pre-existing states, he still continued to refer to "the unavoidable interaction between the objects and the measuring instrument".[58] Bohr's use of such a locution does not commit him to holding that the object possesses inherent states, but allows reference to be made to what were the physical grounds for saying that the ascription of a state to the atomic object was inseparable from the ascription of a state to the measuring tools.

This was, at any rate, Bohr's conclusion, and he proposed that the word "phenomenon" be used exclusively to refer to cases where "unambiguously communicable information"[59] is available, which is information that includes a description of all relevant features of the experimental arrangement. A phenomenon is not an object being subjected to disturbance by measurement. Nor is it an object whose attributes are created by observation owing to the fact that the meaning of these expressions is not well-defined, the impracticability of a causal spatio-temporal description depriving us of the conditions for talking about well-defined objects having definite states. A phenomenon is what we have whenever an object interacts with the measuring instrument, the perception of which is made possible by the observational conditions under which a meaningful ascription of concepts such as momentum and position may be made. To use the words that were Bohr's own when making an allusion to the

Warsaw lecture, "I advocated the application of the word *phenomenon* exclusively to refer to the observation obtained under specified circumstances, including an account of the whole experimental arrangement".[60] I read Bohr to hold that the term does not refer to the entire measuring instrument as such but to be saying that an account of the measuring instrument, that is a specification of whether it is rigid or loosely connected to the experimental frame, is necessary for the meaningful application of one of a pair of conjugate variables in the description of what is observed. In the Warsaw paper he adds, furthermore, that "any measurement in quantum theory can in fact only refer either to a fixation of the initial state or the test of such a prediction, and it is first the combination of measurements of both kinds which constitutes a well-defined phenomenon".[61] Thus, a phenomenon is not what is yielded by a single measurement, a preparation or a detection, but is both. Again Bohr here tries to take the sting out of the EPR paradox.

Naturally enough, the EPR paradox had forced Bohr to change his terminology. The way Bohr subsequently used the word "phenomenon" is different from how he had used it before 1938. Prior to being challenged by Einstein and his collaborators he sometimes spoke of the phenomenon as if it were something over and above the observed object, that is, the object whose properties are determined by observation. In the Como paper and in the other early papers the phenomenon is what is being disturbed when it interacts with the observing instrument. The phenomenon may also be regarded as the subject to which the complementary aspects and properties identifiable in the course of mutually exclusive experimental arrangements may be ascribed. Even Bohr sometimes expressed himself as if atomic phenomena possess properties which constitute classical states but which are disturbed by the measuring interaction in such a way that knowledge of the states of the atomic phenomenon becomes impossible.[62] But it is no great step to go on from here to conclude that behind the experiential manifestations of the atomic object there must exist an object with its own inherent states. Since this was an inference Bohr would resist on any account he had to change his terminology on this point. At the same time I believe that Folse is right when he argues that Bohr himself thought of the phenomenon being disturbed in terms of a phenomenal object whose properties observation determines.[63] No doubt the aim of science was, for Bohr, that of being able to give a description of the atomic object as it appears to us when observed. But a phenomenal object cannot possess, strictly speaking, complementary properties, nor can it be disturbed since it is what is given us in observation. So Bohr's early notion of a phenomenon was indeed incoherent. There was therefore every reason for him to revise earlier formulations supporting a use of the word "phenomenon" according to which a phenomenon could be classically defined in terms of a causal space-time description.

From 1938 onwards he preferred a usage of the word "for comprehension of the effects observed under given experimental conditions", which is equivalent to a usage according to which "any well-defined phenomenon involves a combination of several comparable measurements".[64] The quantum phenomenon is thus characterized by the feature of wholeness since it cannot be subdivided without there being a change in the experimental arrangement, which would be incompatible with there being a manifestation of the very same phenomenon. The description of the phenomenon has to take into account the observational conditions under which it appears and this being so, the description cannot be considered to be one attributing inherent properties to physical objects. This is due to the fact that the interaction between the objects and the measuring instruments "forms an integral part of the phenomena".[65] A consequence of this way of speaking, where the word "phenomenon" refers to a combination of two or more results obtained under specific experimental circumstances, a usage which Bohr now believes is more in accordance with natural language and epistemology, is that phenomena apprehended under various, exclusive experiments are complementary.

> The impossibility of combining phenomena observed under different experimental arrangements into a single classical picture implies that such apparently contradictory phenomena must be regarded as complementary in the sense that, taken together, they exhaust all well-defined knowledge about the atomic objects.[66]

Thus, our entire knowledge of atomic objects can only be described by referring to complementary phenomena each of which, as we have just seen, arises from the interaction between the atomic object and a specific experimental arrangement.

It is quite obvious that Bohr supports the strong meaning condition, as Murdoch calls it.[67] The mere presence of the measuring instrument is only a necessary condition for the meaningful ascribability of position or momentum to the atomic object. What is also needed for a meaningful application, according to Bohr, is that measurements are made. Indeed if no measurements take place there can be no phenomenon to provide us with experience on the basis of which we may ascribe the determinate value of a selected parameter to a particle. The type of measuring instrument determines only the kind of concept which may be meaningfully applied if acts of measurement are actually performed. Nevertheless, when discussing Bohr's semantic argument Murdoch claims that Bohr ought to have adopted the weak meaning condition to the effect that the presence of an appropriate measuring instrument is a sufficient condition of the meaningful ascribability of physical predicates.[68] The reason is, he says, that the strong meaning condition does not go together with the objective-values theory to which the measurement of the value of a certain observable reveals the pre-existing value immediately *before* it has been measured. This theory of measurement is the one which Murdoch believes that Bohr held.

I think Murdoch is right insofar as it is difficult, if not impossible, to subscribe to both views. If it makes any sense to talk about a pre-existing, though relative value, it must also be possible to specify the experimental conditions under which it purportedly makes sense to ascribe the relevant predicate to an object independently of whether an actual measurement takes place or not. But the question is whether the weak meaning condition makes any sense. I fail to see that it does.[69] For if the mere presence of a position-measuring instrument were sufficient for the ascribability of such a predicate to an object, there would be nothing which could determine the identity of the object we were talking about. Is a reference to the experimental set-up sufficient for the meaningful ascription of a position to an atomic object on the moon if the instrument is located on earth? Decidedly not. It is generally assumed that reference to the experimental arrangement can only have significance for the ascribability of properties to particles with a high degree of proximity. Nor do we attribute a specific position to all particles in the vicinity of the experimental set-up. The ascription is to one and only one particle which is non-arbitrarily selected from among all. But this is subject to there being a way of individuating and identifying the particular particle in question. Thus the only thing that can determine which object we are talking about seems to be the actual interaction between the instrument and an object, for it is only in virtue of this interaction that we may individuate an atomic object and thus through that we may ascribe to it the relevant attributes. If, however, the weak meaning condition fails because of the lack of an appropriate individuation principle, the next question is whether or not Bohr held, as Murdoch claims, the pre-existing value theory as his own. If he did, he was then committed to holding that the particle is in a definite state independently of the measuring instrument with which it interacts.

Bohr's way of characterizing atomic phenomena raises, indeed, a very similar question. For if a phenomenon is constituted by effects which can be described unambiguously by taking into account the experimental conditions under which they occur, and if a phenomenon can be characterized as a combination of effects deriving from the interaction between the atomic object and the measuring instrument, is the phenomenon not something different from the atomic object? Well, one might say, since an atomic object is identifiable in the form of complementary phenomena in mutually exclusive experimental conditions it cannot simply be one composite phenomenon. And yet, if an atomic object can manifest itself through its interaction with the measuring instrument in various ways, does this not imply that the atomic object is a reality behind diverse phenomenal appearances? To answer this question affirmatively seems very natural, and this is also what some modern scholars would have us believe Bohr does.

Chapter VIII

1. BOHR AND REALISM

Most of us have from childhood had strongly realistic intuitions in the sense that we believe the physical world to exist independently of whether or not it is observed or can be observed; similarly, we feel quite certain that the properties things have do not depend for their exemplification on whether we cognize them or not. Almost everyone believes that the moon is there without looking to check, and that it would have the mountains it has even if it were not possible to send a space shuttle around it. So facts about the physical world are not something we produce but exist independently of our mental capacities. This view underlies our everyday dealings with the natural world and other human beings, but it has also played a crucial role in the presuppositions underlying most scientific activity. Thus the aim of science has been to produce an objective description of the world to the extent that its nature and structure is undisturbed by human interests, emotions and values. At length, after a great deal of labor, scientists have succeeded in constructing an objective picture of the world according to which physical reality consists of entities which are built out of atoms and the forces at work among them.

Bohr's interpretation of quantum mechanics at the beginning of this century appears to call this realistic picture into question. The kinematic and dynamic concepts do not represent inherent and absolute properties. Such concepts are well-defined only under experimental conditions where the evidence provides us with cognitive grounds for their application. The reason for this is, as will be recalled, that the conditions required for the applicability of classical causal description in space and time cannot be met. Therefore new conditions have to be specified on the basis of which a coherent use of concepts such as position and momentum, as they figure in quantum mechanics, can be established. And these conditions, which are necessary and sufficient for their appropriate use, comprise taking into account both the presence of the measuring instrument and the individual acts of measurement. Bohr also considered the state vector to have no ontological status but merely to be a heuristic device for the calculation of the probability of a specific outcome of the measurement of either position or momentum.[1] Bohr maintained furthermore that the aim of physical science is not so much through it to be able to give an explanation of the intrinsic nature of things as it is to make possible unambiguous communication about experience. These various points all seem to be associated with non-realism in the form of instrumentalism, phenomenalism or subjectivism. But, nonetheless, some scholars have argued that Bohr is not an anti-realist at all but a realist in spite of the fact that many of his utterances seem to run contrary to this view.

What is essential to realism has been characterized in as many ways as there have been authors writing about the subject. However, common to most formulations are two notions to which one seems to be committed in subscribing to realism: (1) the world exists independently of our minds; and (2) truth is a non-epistemic notion; that is, a proposition is not true because it is provable or knowable. This definition squares with Michael Dummett's influential statement of the primary tenet of realism which, together with his statement of the basic claims of anti-realism, has been central to the philosophical debate about semantics for nearly two decades. In Dummett's view the speaker's understanding of the meaning of a statement is equivalent to his knowing the circumstances in which the statement is true and in which it is false. However, a realist and an anti-realist part company when it comes to the concept of truth. Thus, realism is the position according to which reality makes statements true or false in virtue of a correspondence between a statement and certain objective states of affairs, independently of our power to establish which of these values it is (the principle of the transcendence of truth conditions), and according to which any declarative statement is either determinately true or determinately false (the principle of bivalence). Even if it is impossible to produce a basis on which we may ascertain the truth-value of such a statement this does not imply that it does not possess any such value. It always has one. The possession of truth-values has therefore nothing to do with our recognition of the grounds warranting their assignment.

Anti-realism, on the other hand, is defined by Dummett as the position to the effect that a statement possesses a determinate truth-value only if this value can be established (the principle of the immanence of the truth conditions), and that the principle of bivalence is not universally valid. The anti-realist subscribes to an epistemic notion of truth, where a proposition can be claimed true or false in virtue of its assertibility or provability. Thus the anti-realist holds that truth conditions are mind-dependent in the sense that whatever makes a statement true or false has to be in principle cognizable by us. However, such a position does not by itself imply idealism, the position claiming that the world is in reality merely a product of the mind. There is, indeed, a difference between holding that no external world exists independently of our mental capacities in the sense that what is believed to be the external world is a figment of the mind, as the idealist does, and holding that the world does not exist the way it does independently of what can be asserted on the basis of experience. The anti-realist need not hold, as a consequence of his view, that what is real is determined or constituted by the act of cognition. What he is committed to denying is that there exist objects or properties that are in principle unobservable or epistemically inaccessible. He rejects the claim that statements of a given class relate to some reality that exists independently of possible knowledge. But an object may be objectively real, that is independent of the mind, to the extent that it is cognitively accessible as only observed, or in principle observable, objects are. The state of affairs which endows a statement with a truth-value may be mind-independent, but it is a necessary condition for the statement to have a truth-value that it be a state of affairs that is susceptible to human cognitive capacities.

It is therefore reasonable to distinguish between at least two kinds of non-realism: objective anti-realism and subjective anti-realism. (There is at least one further type of non-realism, non-cognitivism, which is a position that not only rejects the validity of the principle of bivalence for a certain class of statements but maintains the validity of the negation of the principle.[2]) The objective anti-realist with respect to statements about physical reality takes, as his point of departure, publicly accessible circumstances when specifying his notion of truth, while the subjective anti-realist takes as his starting point the sensations of the subject or other mental states when saying what truth is. Objective anti-realism is, then, the position which holds that truth is a concept which relates to circumstances whose occurrence or non-occurrence is, in principle, empirically accessible to our cognitive capacities, while subjective anti-realism is the position according to which truth is a concept which relates to circumstances constituted by sense data or circumstances created by our perception. It follows from these definitions that it is only the subjective form, phenomenalism or idealism, which contains the two minimal commitments which are the very reverse of those of realism: (1) no mind-independent world exists, and (2) truth is dependent upon our cognitive faculty. The objective form, however, does not sign away the objectivity of the circumstances which are available for the assertibility of descriptive statements. What it maintains is that we cannot sustain a notion of our descriptive language as one having a content which makes it possible for us to speak about a fixed and objective reality, in virtue of which our statements are determinately true or determinately false, independently of our cognitive means of ascertaining which value it is. This means that the anti-realist only asserts that we cannot distinguish between truth and our capacity to apprehend that in virtue of which the statements are true. He does not necessarily maintain that what makes a statement true or false does not exist independently of the means by which we know its truth-value. The substance of the view of objective anti-realism is then (1) there is an objective, mind-independent world, but (2) truth is related to our cognitive powers. So it is not only realism which includes among its assumptions one making the claim that we operate with a mind-independent world; objective anti-realism does, too.

Following Dummett the quarrel between realists and anti-realists about the physical world may therefore be characterized as a dispute about how judgments, beliefs or statements are invested with a truth-value. Fundamental to this debate then is the distinction the anti-realist makes between decidable and undecidable statements, claiming that the decidable ones are those which are either determinately true or determinately false owing to our possession, in principle, of adequate cognitive means or perceptual evidence by which we might verify or falsify them. In other words, such sentences do have verification-accessible truth conditions. The complementary class of statements is one whose members are undecidable and thus do not have any determinate truth-values, owing to the fact that such sentences have verification-transcendent truth conditions. Yet, in opposition to the anti-realist, the realist would say that even these undecidable sentences have a determinate truth-value; it just happens that we are incapable of finding out which. Nevertheless, both the realist and the objective anti-realist operate with a notion of objectivity. As Crispin Wright has rightly argued, the notion traditionally involves the idea that when it comes to objective matters it makes sense to distinguish, on the one hand, between our beliefs or opinions about a case and, on the other hand, what is, independently of these attitudes, the truth about the case. It is a clear demand for anyone who adheres to objectivity that neither human opinion nor the way this opinion is acquired constitutes truth. Crispin Wright calls the requirement for objectivity the thesis concerning investigation-independence.[3] Thus, the conception of objectivity involves the idea that a certain class of statements has truth-values prior to and independent of any *actual* investigation of the states of affairs the statements denote; that is, independent of the way we investigate such states of affairs or whether we entirely hold back from doing so. With the constraints thus specified it follows that the realist position entails that undecidable, verification-transcendent statements as well as decidable, verification-accessible statements possess investigation-independent truth-values. But the objective anti-realist will, in contrast, only subscribe to the thesis of investigation-independent truth-values with respect to decidable statements. He may do so because he believes that such sentences are endowed with a determinate truth-value since it is always possible to discover what it is that makes them true or false should we look into the matter.

There has been, for a prolonged period, a standing discussion of Bohr's philosophical understanding of quantum mechanics and the ontological assumptions it involves. The problem is that there are very few remarks in Bohr's writings which can be used to settle the question as to whether he believed in atomic objects existing independently of the mind of the observer and, if he did, what he considered their exact nature to be. He never became embroiled in the classical ontological debate about the existence of atomic particles. Most of his philosophical statements about the understanding of atomic phenomena are either of an epistemological or a conceptual nature. So we have to draw many of the ontological consequences of his epistemology and semantics ourselves. In this section we shall see what the arguments are for Bohr being a realist.

The above definition of semantic realism as the conjunction of two principles is the standard one. However, it has recently been argued by several philosophers that bivalence should not necessarily be built into the definition of realism, nor does bivalence follow as a non-trivial consequence of the other half of the standard definition. If this claim is valid, it opens the door to a so far neglected way in which the realist might handle the problem of the specification of the conjugate variables in quantum mechanics. According to Stig Andur Pedersen, the realist may insist that there are certain objective states of affairs which fix the truth-values of statements, as for instance those expressing the indeterminacy relations, such that the function assigning a truth value to a sentence on the basis of such states is uniquely specified semantically, but where that truth function will be a partial one.[4] This implies that the realist must grant the existence of "vague objects", i.e. objects of which only some of the properties are determinate, with others being indeterminate or vague since they possess merely a certain likelihood of occurring. Thus atomic objects will be vague, since they are governed by the indeterminacy relations, which means that there are certain properties that we cannot ascribe with any degree of precision simultaneously. They will possess objective properties, which can be detected experimentally without the properties of the system being changed, while at the same time possessing other, non-objective properties, namely those that cannot be measured without thereby introducing a change in the state of the system.

There is no doubt that the non-bivalent version of realism sketched here is a consistent position; however, its plausibility remains to be argued for. It is consistent insofar as it squares with the above commitments of minimal realism: atomic particles are real, having properties which do not depend on our minds, and it is reality which yields the truth-values; it merely happens that atomic particles are so structured that they sometimes fail to invest proper statements about their properties with any sharp truth-value. The non-bivalentist realist claims that there exist in nature verification-transcendent circumstances which determine the truth-value of a certain sentence; it is merely the case that the truth function which assigns a truth-value to a sentence on the basis of these objective conditions is partial. And this is why sentences about Hermitian observables are not always either true or false.

The non-bivalentist realist holds a position which may be called, following Newton-Smith, "the arrogance response" to the problem of measurement in quantum mechanics. Newton-Smith distinguishes between two kinds of responses a realist might make to the thesis of the underdetermination of data.[5] If you have two rival theories they may be observationally equivalent in the sense that all the evidence we could possibly have wouldn't determine which of the two theories is true. One response would then be to say that one or the other is true but we shall never know which it is. This is what he calls "the ignorance response". It implies the acceptance of the existence of inaccessible facts, i.e., facts which are beyond our power to discover but which make one theory true and the other false. The alternative response, the arrogance response, assumes that there is no matter of fact at issue which conveys a certain truth-value to any undecidable sentence. That is, there are no inaccessible facts and, in consequence, the idea that undecidable sentences are either true or false has to be rejected.

The arrogance response, as it is presented here, indeed holds that there exist objective truth conditions, though they will only partially result in there being determinate truth-values. In that respect this species of realist position certainly differs from an anti-realist position. The anti-realist holds that truth conditions are non-transcendent in the sense that their realization or lack of such has to be characterized in terms of what can be known. However, the arrogant realist and the anti-realist join hands in denying the principle of bivalence for undecidable statements and allowing it for decidable ones. The problem is, therefore, to establish what arguments the arrogant realist may endorse, which the anti-realist cannot, in support of the existence of verification-transcendent truth conditions.

Generally, the bivalentist realist's belief in verification-transcendent truth conditions is tied up with his belief that every statement has determinate truth conditions independent of whether what constitutes their satisfaction is knowable or not, and that every object possesses determinate properties. Many realists will indeed find it hard to give any argument for a realist view if it is maintained that only decidable statements are to be interpreted realistically, i.e., only those statements which allow verification in principle correspond to objective properties, whereas all non-decidable sentences are to be interpreted non-realistically. But this is the wrong way of putting the case. Given the abandonment of bivalence in favor of a position which refuses to ascribe determinate truth-values to certain statements for which no such value can be established, the retention of the idea of verification-transcendent truth conditions seems perfectly arbitrary according to most realists. That retention is dependent upon the possibility of giving sense to the notion that it is not our failure to establish the appropriate truth-value assignments which renders the object indeterminate; on the contrary, it is the intrinsic indeterminacy of the object that is responsible for the epistemic failure. But, one might say, this is metaphorical language which cries out for a literal interpretation. Until the arrogant realist gives it one, it seems sheer dogmatism to insist on the existence of an objective reality which fixes the truth and falsity conditions of a decidable sentence, since the decidability of a sentence is a notion which is defined not in terms of reality but in terms of our knowledge. So the common conclusion is that it is difficult to admit the viability of realism once the universal validity of the principle of bivalence is abandoned.

The non-bivalentist realist is not an objective anti-realist. He endorses verification-transcendent truth conditions in that he maintains that some statements are indecidable only because there exist vague objects, that is objects to which properties cannot be ascribed. However, it is, I believe, possible to submit two arguments, not to be dismissed out of hand, for being a realist although of the non-bivalentist kind. One argument is of a semantical nature in that it draws on the fact that anti-realism allows shifting truth-values since whether or not a statement is true or false depends upon our ability to verify it at a given time. So its truth-value may vary from one time to another. But this is quite unacceptable, it might be objected. A sentence referring to something in the external world cannot have truth conditions which can be expressed only in terms of what it is possible for us to know. The truth conditions for statements about our mental states may be dependent upon our epistemic capacities, but not so those for statements about the physical world. What the realist does in contrast to the objective anti-realist is to insist on stable truth-values, whatever they may be. However, invariable truth-values presuppose the existence of objective, determinate truth conditions.

The other argument is of a metaphysical character. The realist may say that we have reason to believe that there is an objective world, and consequently to believe in the existence of transcendent truth conditions, insofar as there exist causal agents. We believe that something exists if, and only if, it has causal powers. The non-bivalentist realist is a realist insofar as he numbers theoretical entities among the kinds of thing that really exist in the world and which may manifest themselves through observable reactions. But insofar as the properties they possess include indeterminate ones we are prevented from making cause-effect correlations. So a realist need not hold that every theoretical statement is bivalent. He may argue that theories about unobservable entities do not always have to be either true or false.

I have presented here what I take to be a sound version of the realism that states that quantum objects are objectively vague. The idea of vague objects appears to give a coherent ontological account of the problem of measurement, although it does not supply us with any explanation of the problem of non-locality. The next question is then whether Bohr's position can be characterized as non-bivalentist realism. My answer is in the negative. His view is, as I understand it, quite the opposite to that of the non-bivalentist realist. The latter asserts that certain statements about the states of the atomic system are undecidable statements. However, Bohr would say that in quantum mechanics the specification of a measuring instrument and the given measurements constitute the criteria which put us in a position to assert certain statements containing classical concepts on the basis of the realization of these criteria. He would say that statements about the position or momentum of an atomic object are not undecidable, owing to the fact that we have conclusive criteria for the assertibility of any such statement. They are effectively decidable in the sense that, whenever we have produced conditions which show us that the chosen criteria are satisfied, nothing can prove that these criteria on which our judgment is based are inappropriate, although our judgment in itself may be false, of course. This is a consequence of the relativity of the states of a particle, not of the vagueness of such states. Bohr's view is that bivalence only applies to well-formed formulas in the language of quantum mechanics and that ascribing truth-value to sentences about unobservable states of affairs are not well-formed formulas in the language. So what Bohr is talking about is the limit of the applicability of concepts. He is not rejecting bivalence so much as the claim that such sentences are well-formed formulas to which bivalence might be thought to apply.

Henry Folse was one of the first philosophers to maintain that Bohr's philosophy is to be understood realistically in spite of its anti-realistic flavour.

He rightly argues that the notion of complementarity implies that the concept of truth cannot be accounted for as a correspondence between the parameters of the quantum theoretical description and some properties possessed by independently existing entities. Physicists describe in an unambiguous way the phenomenal object as it is revealed by the measuring instrument, he says, not the atomic object. But he rejects the claim that Bohr reached this insight through an epistemological analysis yielding as a result the impossibility of knowledge of the relationship between the observed phenomena and an independent atomic object.[6] Instead Folse argues that Bohr's road to complementarity was the consequence of his attempt to revise the classical view and was triggered by the purely empirical discovery of what is expressed by the quantum postulate. He did not reject realism, Folse says, but merely the classical version of it.

There is, I think, both something right and something wrong in this. Bohr's philosophy is first and foremost a far-reaching epistemological challenge to the metaphysical realism which classical mechanics was generally assumed to support. That this is so can be seen from the fact that, if from nothing else, Bohr continually draws comparisons between quantum mechanics and the theory of relativity. The latter does not feature the quantum of action, so there would be no grounds for making capital of Einstein's theory if it were not for the fact that Bohr thought that it had something in common with quantum mechanics. In his opinion both theories share the same epistemological reorientation in virtue of the conflict existing between their observer-relative basis and the fundamental tenet of metaphysical realism that there exists an observer-independent world and that knowledge of this objective world must be expressed in statements which refer to properties possessed by an object independently of observation. But indeed the controversy was in both cases started by the discovery of the empirical fact of the light constant and the quantum of action, respectively.

Where Folse and I part company is when he concludes that the atomic objects described by quantum mechanics can be said to have a reality quite different from that revealed by their phenomenal appearances and that they have what he calls an independent reality over and above that which becomes manifest in the circumstances in which it is possible to make an assessment of the ascriptions of certain attributes to an observed object. His reading of Bohr is, I believe, not at variance with what Bohr intends. Nor do I think that Folse's reconstruction of the ontology implicitly underlying the notion of complementarity is at variance with his correct analysis of Bohr's view of descriptive knowledge as differing from representationalism. But let us look at the arguments Folse adduces in favor of his suggestion that the framework of complementarity involves a realm of independent objects whose existence is distinct from what the knowing subject is capable of grasping and which produces the phenomenal object.

Folse presents various kinds of arguments which purport to show that complementarity does not have phenomenalism as a consequence, but rather that Bohr was a realist believing in the existence of atomic objects as ontologically independent entities. His first argument is as follows. Although Bohr and the phenomenalists agree that the description of the observation of atomic systems as a form of discontinuous interaction is a consequence of its description being set within a theory which adopts the quantum postulate, Bohr insists, contrary to the phenomenalists, that the quantum postulate is the result of an empirical discovery, forced upon us by the way nature is. The phenomenalists, on the other hand, argue that since atomic objects are not phenomenal objects and are therefore constructions yielded by our theories there may be alternative theories which yield different constructions and which do not presuppose that observations in quantum mechanics involve discontinuous action.[7]

A second argument is this: Complementarity has it that classical concepts refer unambiguously only to properties of the phenomenal object. But since the observing system is given as a phenomenal object and the distinction between the observing system and the observed system is arbitrarily invoked for the purpose of objective description, and since the atomic system, which interacts discontinuously with the observing system, cannot be described in classical terms, the atomic system cannot be identical with the phenomenal object. Therefore, if Bohr had been a phenomenalist, he ought, accordingly, to have asserted that observation is an interaction between a theoretical construction and the measuring instrument, which is absurd.[8] However, this argument goes through only because Folse includes among his premisses ones embodying the assumption that the atomic system is an isolated, independent object having some inherent states which, whatever they may be, change discontinuously in the course of its interaction with the measuring instrument. In that case, and only in that case, is it correct to say that the atomic system is not describable in classical terms but that only the phenomenal object is. But Folse still has to prove that it makes sense, from the perspective of the complementarity thesis, to speak of atomic objects in isolation.

I believe that the idea of discontinuous interaction between the atomic system and the observing system, which plays an essential role in this argument, belongs to Bohr's pre-1935 talk of "disturbing the phenomena by observation" (where "phenomena" refers to the atomic object). Only if atomic objects can be considered to have inherent states possessed independently of any relations to any other object does it make sense to talk about the disturbance of the atomic object through observation and hence, because of the quantum of action, a discontinuous change in its states is brought about by the observational interaction. This might have been Bohr's view until 1935, but subsequently at least he became aware of how misleading the expressions denoting the disturbance of the objects were, since such locutions seem to indicate the existence of classically definable states which might be disturbed discontinuously but not be known empirically.

The third of the arguments goes like this: After 1935 Bohr speaks of "complementary phenomena" and "complementary descriptions", intending by these terms to indicate that two descriptions of different phenomena obtained in different experimental circumstances are complementary if, and only if, they are about one and the same object. For instance, Bohr comments:

> Information regarding the behavior of an atomic object obtained under definite experimental conditions may, however, according to a terminology often used in atomic physics, be adequately characterized as *complementary* to any information about the same object obtained by some other experimental arrangement excluding the fulfillment of the first conditions.[9]

But, according Folse, "the same object" to which Bohr refers cannot be the phenomenal object since it is nothing beyond modes of appearance, and the phenomenal object cannot be spoken of as exhibiting different appearances under differing experimental situations. What is referred to as "the same object" therefore has to be the atomic object which interacts with the measuring instrument and in virtue of which interaction causes the complementary phenomena.[10]

These are the three most important arguments put forward by Folse for Bohr not being a phenomenalist but a realist who asserts the existence of independent, unobservable entities. He is quite right in saying that Bohr's notion of complementarity is incompatible with phenomenalism. But he is wrong insofar as he thinks that this fact entails that atomic objects are some kind of transcendental entity. It is one thing to prove that Bohr was not a phenomenalist, it is quite another to show that he was a realist, believing in the existence of objects which belong to the transphenomenal, noumenal sphere. Bohr was at this point not a Kantian, nor was he a phenomenalist. A person who subscribes to phenomenalism does so by presupposing the truth of (1) an epistemological claim to the effect that we can have knowledge only of perceptible phenomena, and (2) an ontological claim to the effect that physical objects are created by our sensory perception or are constituted by "sense-data". Bohr would certainly deny the latter claim, but there can be no doubt that he would embrace the former. Thus he would oppose the idea that material-object statements like experimental statements are translatable to sense-datum statements, or that the truth of experimental statements is given by the truth of sense-datum statements. Instead he replaced the ontological claim of phenomenalism with the assumption that the physical world is objective in the sense that the content of experience, insofar as what we observe may be given a causal space-time description, exists independently of the mind. One might therefore expect, by introducing the notion of objects underlying the phenomenal objects, as Folse does, that he would be committed to the claim that Bohr held that truth is something which transcends our cognitive powers. But since Bohr maintained that all that we can acquire in the way of scientific knowledge is what is accessible to us through experience as being subsumable under classical concepts, truth, according to him, becomes nothing but a feature of our cognitive capacities.

In considering Bohr's atomic object to be something beyond the realm of phenomenal objects and therefore inaccessible to cognition, Folse involves himself, I think, in an inconsistency. On the one hand he argues that, according to Bohr, the classical concepts of position and momentum apply only to the phenomenal object so "we can use the causal mode to *describe* a phenomenon as an observational interaction in which the phenomenal object has a causal effect on the state of the observing system".[11] Moreover, we cannot use these concepts and the associated conservation principles "to describe the atomic object as the *cause* of the experienced phenomenon". In Folse's construction of Bohr's view the latter uses the concept of causality so that it applies to "the relationship between observed object and observing system as descriptive categories within the whole interaction phenomenon". Any mode of description employing the classical concepts can describe only phenomenal objects. And if one regards any of these modes of description as providing us with an account of a concrete object beyond experience as the "cause" of phenomena, it would lead "epistemology into the pitfalls of representationalism and the notion of a 'real' object possessing at least some properties corresponding to the properties of phenomenal objects".[12] So far, so good. But, on the other hand, when Folse comes to explain more closely what the nature of the independent reality of an atomic object is, he claims that such an object is endowed with the power to interact with other physical systems so as to produce the phenomena we observe, or that it is an "object which interacts with the observing instruments to *cause* the phenomena".[13] This construal is what complementarity requires, Folse concludes. Nevertheless such a way of speaking is highly confusing.

Sometimes Folse speaks of the phenomenal object as the cause of the phenomenon which appears in the interaction with the observing system and sometimes he speaks of the atomic object as the cause. But how can the concept of a cause, the idea of a productive power, be applied to the atomic object if it is only cognitively meaningful to use it for the description of the phenomenal object? This tension is irresolvable so long as the atomic object is regarded as a transphenomenal object which has the power to interact with the observing system producing the phenomenon, but whose properties cannot be described in terms of classical concepts, and so long as the phenomenal object, the attributes of which can be thus characterized, is regarded as what causes the phenomenon we observe.

I think that Folse entirely misses Bohr's point here. He fails to recognize that the observed object, or the phenomenal object as he calls it, may be the independent entity itself, the atomic object, which manifests itself to us through observation, although he sometimes speaks as if the phenomenal object causes the phenomenon through its interaction with the measuring instrument. In fact Folse seems to oscillate between a position to the effect that phenomenal objects are objective in that they have a causal effect on the state of the observing system, and a position to which they are merely subjective by not existing independently of our observation. Folse does not realize that even if an object is as it appears, or rather even if an object cannot in any rational sense be claimed to possess any property whose exemplification it is beyond our cognitive power to establish, it need not be mind-dependent in the sense that its existence has to be a product of the mind. In other words what constitutes the content of experience may still be objective in spite of the fact that the form under which this content is or can be grasped is subjective. Had he seen this, he would not have associated Bohr with the problematic notion of a transcendental reality causing the phenomenal object.

The assumption that atomic objects are transcendental objects involves, at the least, the claim that they are characterized by having certain inherent states which are not accessible to such cognitive powers as we possess. Such transcendental states are states inhering in the atomic object independently of any physical relationship obtaining between it and any other object or independently of our powers of observation. Folse maintains, of course, that Bohr argued that classical descriptive concepts are not – due to the physical fact of the quantum and the empirical reference of these concepts to the phenomenon – sufficiently well-defined to serve as elements in the description of such putative objects. And consequently we have made at least one discovery about such "entities", namely that atomic objects do not possess inherent states as classically defined, and that therefore the possibility of a particle or wave ontology along classical lines is ruled out. But to talk about the atomic object as a reality lying beyond the observed object is to suggest that it has some properties which are such as to transcend our ability to determine or ascribe a certain value to them – besides having such relational properties as render it observable and to which classical concepts apply.

Indeed the problem which Folse faced was this: if Bohr is not a phenomenalist, as he rightly argues, but assigns to atomic objects the status had by independent features of physical reality, and if Bohr asserts further that quantum mechanical statements provide us with a non-reductive description of these objects, then Bohr cannot be, it seems, anything but a mere Kantian in disguise in virtue of making what seems like a distinction between the phenomenal object and the noumenal object. But this conclusion follows only if it can be assumed that the predication of a certain property to an object has to be characterized as the ascription of an inherent state to the object. For then this state cannot belong to the atomic object itself, and it must therefore belong to something else, namely, the phenomenal object. However, the way Folse explicates Bohr's intention by letting him insert the phenomenal object between the observer and the atomic object creates problems not only for his interpretation just mentioned. If Bohr were the realist Folse takes him to have been, he would have had to assume that atomic objects possess definite inherent states whose nature it is impossible to ascertain empirically, since no possible experience would carry us beyond the phenomenal object. Yet, nowhere do we find anything which indicates that Bohr subscribed to such a view. The fact is that Bohr did not distinguish between the atomic object as a transcendental object and as a phenomenal object.

Folse takes Bohr to be an "entity realist" in the sense that Bohr believed objects describable by quantum mechanics to be caused by real unobservable entities existing independently of our knowledge of them.[14] This view might seem to be supported by a long passage from 1929. Here Bohr says:

> Natural phenomena, as experienced through the medium of our senses, often appear to be extremely variable and unstable. To explain this, it has been assumed, since early times, that the phenomena arise from the combined action and interplay of a large number of minute particles, the so-called atoms, which are themselves unchangeable and stable, but which, owing to their smallness, escape immediate perception. Quite apart from the fundamental question of whether we are justified in demanding visualizable pictures in fields which lie outside of the reach of our senses, the atomic theory was originally of necessity of a hypothetical character; and, since it was believed that a direct insight into the world of atoms would, from the very nature of the matter, never be possible, one had to assume that the atomic theory would always retain this character. However, what has happened in so many other fields has happened also here; because of the development of observational technique, the limit of possible observations has continually been shifted. We need only think of the insight into the structure of the universe which we have gained by the aid of the telescope and the spectroscope, or of the knowledge of the finer structure of organisms which we owe to the microscope. Similarly, the extraordinary development in the methods of experimental physics has made known to us a large number of phenomena which in a direct way inform us of the motions of atoms and of their number. We are aware even of phenomena which with certainty may be assumed to arise from the action of a single atom. However, at the same time as *every doubt regarding the reality of atoms has been removed* and as we have gained a detailed knowledge of the inner structure of atoms, we have been reminded in an instructive manner of the natural limitations of our forms of perception.[15]

In another essay from the same year Bohr expresses a similar attitude to the reality of atoms: "We know now, it is true, that the often expressed skepticism with regard to the reality of atoms was exaggerated; for, indeed, the wonderful development of the art of experimentation has enabled us to study the effects of individual atoms".[16] This might be interpreted as a realist position.

It is, however, quite obvious from these two passages that Bohr does not say anything about the nature of the reality belonging to the atom. He does not even say that atoms are real entities existing apart from their phenomenological appearances. What he seems to refer to is the hypothetical character which the theory of the atom once had when it was put forward without any observational evidence to explain the variance of chemical phenomena, but which has subsequently been removed due to the fact that scientists, at the time he was writing, had acquired strong empirical evidence supporting the assumption that every chemical element is made up of atoms. The doubt surrounding atoms has diminished because there is a vast amount of experimental evidence for their existence. Bohr says that atoms are real because the physicists have proved them to be so. With that interpretation in mind his remarks concerning the indubitable reality of the atomic world may be construed anti-realistically just as well as realistically. Instead of being a clue to Bohr's ontology these statements must be understood, I believe, as a reaction to the denial in the late nineteenth century, by Ernst Mach and other chemists, of the existence of atoms.[17] Mach was a phenomenalist par excellence, which is to say, a subjective anti-realist, who maintained that both physical and mental phenomena are reducible to certain basic elements of perception, sense data (*Empfindungen*), which are claimed to be ontically neutral. Only what is analyzable in terms of these elements satisfies the criteria for intelligibility, and the aim of science is thus to describe the interrelationship between these elements.[18] So the notion of the atom's constituting a theoretical entity need not amount to giving it an ontologically independent status but merely serves an instrumentalistic or heuristic purpose. However, this position definitely does not represent Bohr's view, nor was it Høffding's. But this does not imply that they have sold their souls to the opposing camp. It might be contended that if the two passages from 1929 are construed as a defense for realism, it becomes inexplicable why Bohr only once stresses the reality of the atoms, viz. in 1929, at a time when Machian phenomenalism might still be fresh in the memory of scientists. Thus what Bohr intended by these remarks was to dissociate complementarity from phenomenalism rather than to associate it with realism.

In an interesting exchange of letters between Max Born and Bohr in 1953 we actually see that Bohr rejected the idea that the atomic system can be thought of as a reality behind the phenomena, but strangely enough Folse takes this correspondence to support his reconstruction.[19] Born had written in a paper which he had sent to Bohr that he "disliked thoroughly" the position of those physicists and philosophers who think that the aim of science is a "purely observational description in which one does not imagine what there is behind the phenomenon".[20] Bohr's first response to the suggestion of something behind the phenomena was strongly negative.

Indeed, it is difficult for me to associate any meaning with the question of what is behind the phenomena, beyond the correspondence features of the formalism which itself represents a mathematical generalization of the classical physical theories permitting, within its scope, predictions of all well defined observations which can be obtained by any conceivable experimental arrangement.[21]

In his reply Born explained what he meant by "behind the phenomenon":

What I meant by "behind the phenomena" is in mathematical language just "invariants" in the most general sense of the word. The various aspects of phenomena which we consider in quantum mechanics have also a theory of "invariants", or in less learned language, common features which do not depend on the aspect, and it is this which I would like to preserve as something beyond our direct experience. ... If one does not accept such a standpoint, it appears to me that one accepts a hyper-subjective or solipsistic standpoint, and that one resigns oneself to answering any question about why one is investigating the world at all.[22]

It is this definition of a reality behind the phenomena in terms of the mathematical concept of invariance that Bohr embraced in the following letter:

I owe you, of course, an apology for my remarks as to the question of what is behind the phenomena. ... As you express your views in the letter, I agree entirely and had myself the same attitude when in my letter I spoke of a consistent abstract generalization of classical mechanics and electrodynamics interpretable only on correspondence lines.[23]

However, it is difficult to see why Folse thinks this agreement confers evidence on his interpretation of Bohr as one who believed in an ontology of unobservable entities invested with the power to interact with physical system to cause the phenomena we observe. As I understand his reply, Bohr was saying that "a consistent abstract generalization of classical mechanics and electrodynamics", which the quantum theory is, gives us the lawful connections, or if you like invariance, between the phenomena we can observe. This is all there is about the reality behind phenomena. The quantum phenomena are given empirically, and they are objectively real in the sense that they are manifested to us in the context of lawful connections. The lawful connection between the atomic phenomena is our grounds for believing that these phenomena are connected by what is not observed, i.e., by atomic objects existing between preparation and detection. Consequently, atomic objects are real insofar as they contribute to the creation of such lawful connections.

Indeed, one may ask what the middle ground between idealism and realism is? If the quantum system has no reality apart from the phenomena in which it is observed, then the quantum system *is* created through the process of observation and Bohr is the idealist various interpreters have claimed him to be. If the electron does have some kind of reality apart from the phenomenal appearances it manifests in observational interactions, then there must be *something* which exists distinct from the phenomena. This is to assume that there must *be* an independent object. A problem here is that Bohr denied outright the intelligibility of ascribing reality to the transphenomenal object, and he did not merely claim that whatever is behind the phenomena cannot be described as possessing properties to which classical concepts would refer. Kant, however, did believe in the reality of a transphenomenal realm. What he denied was the ability of humans to acquire scientific knowledge of transcendental objects. But neither Bohr nor Høffding stood with Kant on this; both refused to accept the reality of the transphenomenal realm. In my opinion, every analysis involving what is the purportedly realistic commitments entailed by Bohr's philosophy and with which we have been confronted above is deficient, and I believe that we will get a better understanding of the issues as to whether Bohr believed complementarity involves further ontological requirements than those which flow from the semantical argument for the indefinability thesis if we look at what Bohr was taught by Høffding about reality and objectivity.

2. HØFFDING ON REALITY

When reading *Den menneskelige Tanke* for the first time one gets a powerful impression that realistic descriptions of scientific knowledge are being entirely dismissed. Høffding held that we ourselves produce the truth whenever we discover the principles under which we may subsume our perceptions. He embraced a very strong principle, one which he termed the "principle of experience", whose import was that experience sets the bounds to truth and knowledge. Reality cannot be talked about intelligibly as existing independently of our capacities of cognition. As he says, reality is the actual truth. Thus reality is what is true, and the description of what is actually the case is the one which yields us the greatest possible number of lawful connections in experience, and such connections are established whenever a causal description which is in harmony with a particular theory can be given of the phenomena involved.

Furthermore, the formal conditions of truth are linked to the existence of consistent connections between various theories. This means that what is true is what is ultimately justifiable on the basis of a theoretical description of our experience as law-governed. So, according to Høffding, truth does not consist of an agreement or a correspondence between statements and a certain absolute realm of facts, because any epistemic access to such a relation is denied us. In this vein, too, he opposes the notion that things may be ascribed properties which cannot manifest themselves in relation to other things and are thus unknowable.

The anti-realist position, as we have seen, squares with much of what Høffding was saying. Yet, Høffding regarded himself as a realist. As early as 1882 he wrote, "The truth and justification of scientific realism will depend on whether it really expresses the only principle which can bring about unity and harmony in our cognition and whether it is consistently confirmed as science progresses".[24] He defined realism as the position maintaining the principle of natural causes. To him the essential difference between idealism and realism lies in their different ways of accounting for our experience. Idealism, on the one hand, accounts for the phenomena of nature by principles drawn from outside the realm of the natural world, namely by claiming that the particulars are produced by the mind of the knower; realism, on the other hand, considers the particulars as given, accounting for them with the aid of principles drawn from the natural world, that is, nature is to be explained by nature itself, as Høffding put it. This means that the particulars of nature are caused by other particulars of nature. His conclusion was, then, that scientific realism is the only view which establishes unity and harmony between the spirit of science and a general philosophy. However, no matter how Høffding characterized himself he denied quite explicitly the existence of a reality behind the phenomena: he rejected any Kantian absolute, an existent "Ding an Sich" or a Lockean substance, "I know not what", independent of any possible cognition, as well as arguing against a static concept of truth according to which truth consists of a correspondence between certain cognitive beliefs and the existence of absolute, unobservable particulars. So Høffding was apparently an anti-realist, in the sense in which the term has been defined above.

However, a more careful reading of Høffding's arguments reveals that he is what I have defined as an objective anti-realist rather than a subjective anti-realist. He maintains, for instance, that most items are not brought into being through the processes of thought, i.e., the particulars of nature stand thus in contrast to the particulars of mind, because the concept of causation applies neither to the ontic relationship of mind and body nor to the epistemic relationship between subject and object. Instead, as it may be recalled, he held that the items of perception are immediately given, they are not produced by us. They appear to the mind as wholes and do not belong to consciousness itself. It is physical phenomena which are experienced: they are *particulars* to be experienced by us, and are thus the object of experience, and they form part of reality insofar as they are judged to be causally connected with other physical phenomena. Thus it is not simply phenomena as such which are real, although they are immediately given as the content of experience, but it is the presence of the causal relationship obtaining between them, as it may be describable in terms of scientific theories, which makes them truly mind-independent entities.

In classical mechanics it has been possible to connect various phenomena by assuming that the physical system still has a well-defined state when unobserved, and accordingly observable properties are attributed to the system even when it is not observed. Confronted with this fact Høffding would probably say that since the properties which we attribute to a system in its classical state are observational properties they may also be part of an actually unobserved reality as long as the theory prescribes such well-defined states in its efforts to establish a causal connection between diverse phenomena. Høffding did not deny the existence of unobserved entities. Beliefs in unobserved items are cognitively justified only as long as these particulars are assumed to have observable properties. What he claimed was that what is for the mind, the given, is what is observable, and the phenomena which in this way can be known constitute reality insofar as they can be described as contributing to the network of causal connections. So what he rejected was the intelligibility of transcendental entities with unobservable properties on the grounds that we have *eo ipso* no empirical warrant for such a belief. He rejected unknowable but not unobserved entities.

In developing Høffding's view it might said that one of its consequences is that a descriptive statement is not true or false in virtue of circumstances whose realization is one that is indifferent to the scope of our cognitive powers. For him the truth of such statements would consist in the satisfaction of certain observable conditions which serve to justify our belief in their truth. And these conditions can so serve if the statement about a particular experience can be coherently connected with other statements which together express the claim that there are continuous connections linking our perceptions. Indeed, knowledge of the truth of these other statements would have to be taken for granted. So Høffding would probably say that a statement is true if, and only if, there is a situation which it is in principle possible for us to observe and that this situation, if observed, would definitely enjoin on us the acceptance of the statement on the basis of our knowledge of the truth-values of other statements. So, the assertibility conditions to which Høffding subscribes contain two components, an empirical one as well as a pragmatic one.

This account of some of the implications of Høffding's epistemology is in accordance with Høffding's thesis that thought is both greater and smaller than reality, i.e., that thought both extends beyond and underdetermines reality. Pushing back the frontiers of knowledge consists in the elimination of certain possibilities in the sense that alternative ideas and hypotheses are ultimately refuted – in this sense reality is smaller in scope than thought about it. Conversely, reality is richer in the sense that it embodies the potential for experience not yet had. Nevertheless, according to Høffding, there exists an incommensurability or an irrational relation between thought and reality. This means that the relationship between thought and reality defies exhaustive description. In consequence Høffding would probably claim that we cannot speak meaningfully about the relation between language and reality. We can merely lay bare the conditions for describing that reality.

When Høffding claimed that the relation between thought and reality is incommensurable or irrational he meant that the knowing subject cannot compare its beliefs with an independent reality as the naive realist suggests. In Høffding's terminology, however, naive realism is the doctrine that what is true and what it is true of, correspond to each other. Because of this "shortfall" the progress of science has to be accounted for in terms of what Høffding called constructive realism. He says, for instance:

> In many ways thought must, as the history of science shows, transform the experienced and immediately given in order to prove the truth of a belief about reality. Naive, secure realism has a concept of reality different from that of constructive realism which seeks to work out a defensible concept of reality and which it is up to natural science, history and philosophy to flesh out. Slowly but surely human thought moves in this direction.[25]

To this he adds that there will always be certain items from which reflection will take its departure in its efforts to deploy the criterion of reality. As he puts it, "Not all items remain on a par; the reflective consciousness must start by presuming the reality of certain items and test others with respect to these".[26] Our thought needs firm ground, something which, *ex hypothesi*, is beyond doubt, then, later on, the mind may subject this hypothesis to critical examination if another kind of purchase on reality has been achieved at that point. Thus, according to Høffding, the criterion of reality presupposes in every case that the existence of certain phenomena is taken for granted. And through the retraction of some claims and the accession of others cognition approximates the truth to an ever-greater degree.

Constructive realism, which is what Høffding calls his own view, is based on a dynamic concept of truth. A principle is true, according to constructive realism, if with its aid we can advance in our understanding, and if it is something which presents itself whenever the reflective consciousness discovers regular connections in the process of ordering phenomena. Naive realism, on the other hand, is built up around the static concept of truth. Recall, moreover, that Høffding had two arguments for the repudiation of the static concept of truth. First, it contains a contradiction because it calls for a correspondence between the belief of the subject existing in itself and objects existing in themselves, but cognition is always interaction between the thoughts and the phenomena, be these either physical or psychical. Second, reality in itself is not capable of being an object of cognition, thus it is impossible for the mind, the reflective consciousness, to compare it with thoughts. The dynamic concept of truth is therefore the only possible one.

Certainly, what has just been said proves that Høffding subscribed to an epistemic notion of truth, one of the commitments of anti-realism. But nevertheless he did not accept the other condition for being a subjective anti-realist: the claim that nothing exists independently of the mind. In fact he rejects it by embracing one of the commitments which is central to the realist position: belief in the existence of an objective, mind-independent reality. Thus his constructive realism comprises one commitment which allegedly belongs to the anti-realistic position and another which is traditionally taken to be part of the realistic position. This point of view is what I have called objective anti-realism. Høffding ruled out metaphysical realism, which he calls naive realism because of its involvement with the correspondence theory of truth and with a substance-property ontology. Instead he argued for constructive realism, holding that truth is an epistemic concept but that reality exists independently of the experience in the form of items subjected to laws. Høffding's view seems in certain respects similar to the internal realism of, for instance, Hilary Putnam and Brian Ellis, notwithstanding the historical differences of expression.

Constructive realism, as it is presented to us by Høffding, involves a strong element of holism, which is alien to naive realism. To see how, let us consider his notion of a thing and its properties. Høffding maintains that a thing or an object is experienced as an immediately given whole. What he has in mind may perhaps be illustrated by the following example. Take for instance the perception of a tree. We perceive it as an immediately given whole, we don't see it as consisting of parts or as a collection of various properties. The shape of the trunk, its thickness, its color, the form of the branches, the leaves and their colors, its bulk and maybe its blossoms, are all properties of the tree which are united into one and the same image. They are not perceived as separate items but as belonging to the tree as a totality. The tree itself is perceived as *one* item. Indeed, we can concentrate our awareness on this or that part of the tree, but it will still be this or that part which is given to us as parts of an experienced totality. But Høffding would certainly add that the tree is not merely an experienced whole, it is objectively given because, besides its observed qualities, it also has many qualities which at any given time are unobserved. These unobserved qualities also contribute to the tree's being a conceptual totality as distinct from being experienced as such.

However, Høffding emphasizes that the tree as an immediately experienced whole or totality is known to us merely through its properties, which are nothing more than its relations to other things. This sounds strange indeed, since how can a tree be experienced as a totality and at the same time only be known through its properties? Høffding's answer would run something like this: Physical things are as they pretend to be in observational situations, their nature is a function of the way in which they manifest themselves in various experiential circumstances. Physical objects cannot be things behind or over and above our possible experiences, entirely independent of our capability of having cognitive access. As an item of immediate experience we see the tree as a whole, in a synthesis, but when we by reflection analyze our experience of the tree we discover that it consists of various analyzable elements.

The reality of the tree may be taken for granted, which, of course, it is in everyday life. But what about theoretical entities such as atoms? Høffding did not reject the reality of atoms *tout court*. What he maintained was that the concept of atoms like the concept of space and the concepts of natural kinds do not refer to anything over and above what is observable. In fact, as we have seen, he accused Kant of not being critical enough when postulating noumenal objects behind the phenomenal appearances of objects in space and time. According to Høffding, there is no substance supporting a spatial or atomic entity on the basis of which the phenomena are given in the observational interaction of man. So if atoms exist they are related to phenomenal objects, which means that the reality of atoms has to be understood with respect to their various empirical manifestations.

Høffding's stand on atoms may be put as follows: The concept of the atom has gradually been developed in order to describe a variety of phenomena and interconnections between these phenomena as a manifestation of the atom's properties. In order to describe the regularity and lawfulness yielded by the results of various experiments and observations scientists have invented the notion of an atom. But the implication does not follow that this idea stands for something behind the various phenomenal manifestations supporting the claims made about such an object. Here Høffding would not distinguish between the manifestations and that which is manifested in appearances. Atoms are real only to the extent that they manifest themselves in lawful connections. They exist as a result of the continuous connections obtaining between observable phenomena. The properties which are attributed to the atom on the basis of their observational manifestations are those which together constitute the atom. But, at the same time, Høffding would certainly say that atoms are nothing but the sum of their possible observational manifestations.

It is my contention that the ideas with which we have been presented in this section also formed parts of Bohr's view on the reality of atoms.

3. OBJECTIVE ANTI-REALISM

So far I have argued that Høffding was an anti-realist although of the objective kind. He denied both the correspondence theory of truth and a substance/property ontology, which together involve a belief in transcendent truth conditions. It is now time to see whether Bohr's philosophy and the notion of complementarity can be understood as a form of the anti-realism which I am attributing to Høffding. I think it can. But where Høffding derived his ontology from an epistemic account of the conditions for possessing objective knowledge of nature, the mature Bohr went further and drew his ontology from a semantic analysis of the conditions for unambiguous communication in quantum mechanics.

If we look at each of Bohr's sayings in isolation there seems at first blush to be a strong tension between the realist and non-realist features built into his philosophy. But taking them into consideration collectively, Bohr seems neither to be a realist nor an idealist in the sense in which I have defined these positions. This was also how he looked upon his own position. Strangely enough, the passage which is perhaps most central to an understanding of how Bohr considered his own philosophy with respect to various philosophical doctrines is not, with Henry Folse as the only exception as far as I am aware, referred to by any of the scholars who claim that he was an idealist or a realist. In 1954 when talking about the epistemological lesson learned from the development of physical science he said,

> In return for the renunciation of accustomed demands on explanation, it offers a logical means of comprehending wider fields of experience, necessitating proper attention to the placing of the object-subject separation. Since, in philosophical literature, reference is sometimes made to different levels of objectivity or subjectivity or even of reality, it may be stressed that the notion of an ultimate subject as well as conceptions as realism and idealism find no place in objective description as we have defined it; but this circumstance does not imply any limitation of the scope of the enquiry with which we are concerned.[27]

And we may add that neither is the notion of an "ultimate object" consonant with what he understands by an objective description.

Thus, on the one hand, Bohr denies the realist idea that the aim of quantum mechanics (or of any other physical theory) is to explain the phenomena in terms of an underlying, hypothetical reality. As he says in one of many related contexts, "In our description of nature the purpose is not to disclose the real essence of the phenomena but only to track down, so far as it is possible, relations between the manifold aspects of our experience".[28] The state function for the electron in "free space" does not "denote" or "represent" the independently real electron as it exists between the phenomena. In other words, Bohr dismisses the claim that theoretical terms like "electron" denote real unobservable entities of which we cannot acquire any observational evidence, and which constitute the conditions for the assertion of their reality. Consequently truth, according to Bohr, relates to experience, and not to a putative reality lying behind phenomena. But, on the other hand, a description of our sensory experience requires that a clear distinction be made between the subject, having the experience, and the object, giving the experience its content, in order for it to be objective and unambiguously communicable to others. As he says about the fields of experience to which the notions of relativity and complementarity apply, "The decisive point is that in neither case does the appropriate widening of our conceptual framework imply any appeal to the observing subject, which would hinder unambiguous communication of experience".[29] Bohr's concept of objectivity implies that only experience which is unambiguously communicable by means of classical concepts, that is by being described independently of any explicit reference to the individual observer, constitutes the sphere of reality. Bohr certainly intends to assert that the view of complementarity (and relativity as well) is different from both realism and idealism – some intermediate position, perhaps? As I read Bohr's account, his view is that the realist strives for more than he can have, while the idealist or phenomenalist wants less than he can get. Bohr rejects, on the one hand, the notion that our imaginative powers may be superior to our cognitive capacities, that truth may intelligibly transcend what can be asserted on the basis of experience. On the other hand, he dismisses the idea that the physical world in itself is but a creation of the mind or a construction of the cognitive capacity. But the problem remains as to how such a position, which I have called objective anti-realism, can be articulated in detail.

The serious question is, indeed, whether or not the objective anti-realism which I attribute to Bohr and Høffding is in fact a coherent position. Is it possible at one and the same time to maintain the existence of a mind-independent reality and the cognitive dependence of truth? I believe that it is, and to show why let me illustrate the thesis with an example taken from outside the realm of quantum mechanics. With respect to the reality of material objects one may adopt either a realist or an anti-realist position. It may be claimed that any descriptive statement which contains a reference to a material object has a determinate truth-value independently of whether we possess any procedure in virtue of which we can assert that such a statement is either true or false. Similarly, our understanding of material-object statements is not related to conditions which are such that if we know them we are in principle able to determine the value of a given statement. If a person makes this assumption, he or she is a realist with respect to physical reality. However, the various forms of non-realism with respect to the reality of the physical world deny that all material-object statements need be either true or false as well as rejecting the claim that our understanding of these statements is based on conditions, knowledge of which does not allow us to determine their truth-value.

The conditions which have to be present in order for an anti-realist to have the strongest possible grounds for asserting that material-object statements have a determinate truth-value are normally assumed to be those given in our perceptual experience. But, as Michael Dummett has rightly argued, the anti-realist may take either a reductive stand or a non-reductive stand towards statements of the given class (in this case the material-object statements), and may either hold that the meanings of that class of statements must be accounted for in terms of the meanings of the reductive class (the sense-datum statements), or deny the intelligibility of such a reduction.[30] Thus the reductive anti-realist, the phenomenalist, would assume that the meanings of material-object statements can only be explained when they are translated into statements concerning sense-data; while the outright anti-realist would argue that the reductive class of sense-datum statements cannot characterize our perceptual experience, the description of which is entirely saturated by a material-object vocabulary. But still, being an anti-realist, the latter would maintain that we possess no notion of truth which transcends our capacity to recognize material-object statements as true.

An outright anti-realist need not oppose the notion of an objective reality: what he claims is that the notion of the truth of a material-object statement is derived from what constitutes the strongest possible grounds for asserting it. These grounds may be physical in origin, hence objective, although any specification of them has to relate to our perceptual powers. So for the objective anti-realist any decidable statement containing a material-object vocabulary refers to objective matters, though whatever truth-value it has is determined in relation to our powers of observation. This means that the states of the world which make us attribute a truth-value to a decidable statement must be detectable by us in order to count as truth-conferring conditions. But, as long as our experience is described in terms of material-object statements which serve as reports of observation, such decidable statements are rightly claimed to be concerned with the objectively real world. That is, all statements about physical reality whose truth-values can be established by means that do not transcend our cognitive powers are about genuinely factual matters.

There is here (to get the view of objective anti-realism straight) a clear analogy with other areas in which anti-realism may be maintained. A person who embraces such a position with respect to the past holds that a past-tensed statement about a physical event is either true or false only if there is strong evidence for asserting or denying it. Such evidence may take the form of a memory. But this fact does not entail that the past-tensed statement refers to the psychological states constituting that memory. The anti-realist may either hold or reject the idea that the meanings of statements about the past can be reduced to the meanings of present-tensed statements about certain psychological states, and consequently the assumption that the past is or is not created by present mental activities. Of course, the objective anti-realist maintains, contrary to the subjective anti-realist, that the translation of any past-tensed statement to present-tensed statement is impossible, that the former type of statements refers to past states of affairs, either physical or psychological, and so therefore that the past is not a mental construction of present memories, nor is it a construction drawing on physical traces and records. Nevertheless, he still commits himself to the supposition that the truth of a past-tensed statement consists in the availability of grounds for holding it true.

Now, Bohr regards the vocabulary of material objects as part of our ordinary common-sense language, which he believes is evolved for the very expedient of expressing the forms of cognition deeply entrenched in our thought, and that its use is thus rendered necessary for the objective description of experience and unambiguous communication. He would resist as unintelligible every idea of translating material-object statements into sense-datum statements in order to understand the meanings of the former through those of the latter. If Bohr is an anti-realist, he is not one of the reductive kind. Descriptions of our everyday experience are couched in the ordinary language which serves to express the conceptual framework presupposed when reference is made to material objects figuring in causal contexts. This framework is in Bohr's view the precondition for any distinction between what is objective and what is subjective, and thus ordinary language, including its material-object vocabulary, emerges as the only language in which we may make unambiguous and objective statements about our experience.

The philosophical vindication of objective anti-realism is therefore related to the possibility of establishing what Bohr calls "unambiguous communication".

As he puts it, "Every scientist is constantly confronted with the problem of objective description of experience, by which we mean unambiguous communication".[31] Bohr holds further that the tool for the formulation of intelligible and objective descriptions, even in science, is "plain language which serves the needs of practical life and social intercourse", supplemented with terminological refinements. Or as he also puts it, "All account of physical experience is, of course, ultimately based on common language, adapted to orientation in our surroundings and to tracing relationships between cause and effect".[32] But what makes a quantum mechanical description unambiguous? In Bohr's answer to this question lies the core of his legacy, and it may be put as follows: in order for a communication to be unambiguous, and hence objective, the use of descriptive terms must be related to observable situations. This is what Bohr means by saying that "the unambiguous account of proper quantum phenomena must, in principle, include a description of all relevant features of the experimental arrangement".[33] The experimental arrangement, including the recording of observations, in its totality defines the conditions which have to obtain in order for the classical descriptive concepts to be meaningfully applicable. But the specifications concerning the measuring instrument and the information acquired through its use serve as the conditions for an unambiguous application of classical concepts only if the description of the experimental arrangement and the recording of observations are communicated in the everyday language supplemented with the appropriate terminology of physics. "This is a clear logical demand", Bohr says, "since the very word 'experiment' refers to a situation where we can tell others what we have done and what we have learned".[34] But this is not the only reason. Everyday language is also a necessary condition for unambiguous communication because of the fact that by describing the experiment in the common material-object language, one of whose very functions it is to express the relationship between cause and effect, we are able to pay "proper attention to the placing of the object-subject separation" that is necessary for unambiguous communication of our experience.[35]

As early as 1938 Bohr seems to have grasped this entire argument.

> We must, on the one hand, realize that the aim of every physical experiment – to gain knowledge under reproducible and communicable conditions – leaves us no choice but to use everyday concepts, perhaps refined by the terminology of classical physics, not only in all accounts of the constructions and manipulation of the measuring instruments but also in the description of the actual experimental results. One the other hand, it is equally important to understand that just this circumstance implies that no result of an experiment concerning a phenomenon which, in principle, lies outside the range of classical physics can be interpreted as giving information about independent properties of the objects, but is inherently connected with a definite situation in the description of which the measuring instruments interacting with the objects also enter essentially.[36]

So Bohr's reasons for connecting the unambiguous communication of quantum phenomena to the experimental arrangement rest on the assumption that such recognizable situations as the particular outcomes of a specific measurement determine, at least in part, the use of classical descriptive concepts and the ascription of a determinate truth-value to statements predicating a certain property of the atomic object. For these situations in which evidence is available constitute, it would seem, the only ones in which we can communicate intelligibly to other speakers about our claims, whereupon they can confirm whether or not we are using the descriptive terms correctly. Ordinary language is a tool geared to the making of statements descriptive of our experience and it is meaningful only within such limits as experience can justify. If this is Bohr's argument, as I believe it to be, he seems to be anticipating Dummett's two central semantic arguments for being an anti-realist.[37]

The first of Dummett's arguments is to the effect that our understanding of the meaning of a declarative statement, and hence our knowledge of the conditions under which it is true or false, is entirely determined by and determines the use we make of the sentence in observable situations. Because of this there can be no features connected to the meaning of a sentence which transcend possible use, since other posited features would entail that there exists something about the meaning of a sentence which is incommunicable. The second argument is to the effect that we as competent speakers of a language learn that language through the association of the use of certain sentences with the occurrence or non-occurrence of certain recognizable situations, and that it is only on the basis of whether we are able to use sentences correctly in such situations that other users of the language can decide whether we have grasped the meaning of the sentences. It would therefore be incomprehensible how features of the meaning of a sentence which did not rest on such observable situations could be learned and be communicated by one user of the language to another. Bohr would undoubtedly have been able to subscribe to these two arguments of Dummett's in favor of semantic anti-realism if he had been acquainted with them, and would have considered the latter's formulations as more precise and more general than his own.

As mentioned above, Bohr very often stresses the fact that scientific theories do not reveal the constitution of nature in a way which goes behind and beyond experience. I take this to mean that theoretical statements are not literally true descriptions since such statements could only be assigned truth conditions which transcend our cognitive powers, and we have no cognitive warrants for so doing. In Bohr's own words when talking about the non-commutable algebra of quantum theory, "Owing to the very character of such mathematical abstractions, the formalism does not allow pictorial interpretation on accustomed lines, but aims directly at establishing relations between observations obtained under well-defined conditions".[38] Thus, quantum theory with its advanced mathematical formalism is an abstraction which has no direct physical content; it serves only to describe the sort of experience which is to had in the appropriate circumstances. If I am right in my contention, this would indicate that Bohr regards scientific laws as instructions for descriptions instead of as contributing directly to the description of the world as it is. Thus nomological laws do not have a descriptive content themselves but are linguistic rules for descriptions.

They determine how a competent speaker of scientific language has to formulate particular descriptive statements in order to communicate unambiguously about any specific experience.

Moreover, the aim of scientific theories is to help us predict new phenomena by synthesizing our knowledge of previous experience in an objective way. As Bohr says,

> The main point to realize is that all knowledge presents itself within a conceptual framework adapted to account for previous experience and that any such frame may prove too narrow to comprehend new experiences.[39]

He then adds, "When speaking of a conceptual framework, we refer merely to the unambiguous logical representation of relations between experiences".[40] This might be understood to mean that answers to ontological questions would always be relative to a theoretical construction based on experience. Bohr would then be endorsing the same kind of ontological relativism as Willard v. Orman Quine, Paul Feyerabend and Thomas Kuhn are seen to do.[41]

If science has only a relative ontological foundation, it cannot be its task to generate knowledge of some fundamental mechanism or entities causing the phenomena. Science does not discover something behind the phenomena because there is nothing there. Methodologically scientific theories give us an economical organization of the phenomena observed in the laboratories and in the world generally. Scientific theories are merely very effective instruments for predicting future experiences on the basis of past experience and thereby giving us a better appreciation of the courses of action open to us.

In Bohr's eyes there is, in certain respects, some truth in this claim. Scientific theories are tools for ordering our experience and making predictions of new experience and their value consists in their ability to do this work. Yet, there is a fundamental difference between Bohr and someone who supports ontological relativism. According to the latter it is in principle possible to construct two mutually inconsistent theories by using different concepts so that these theories considered in isolation describe all empirical facts. For Bohr, however, there is no alternative framework to one using the classical concepts, which are refinements of those of the common-sense conception of the world. Classical concepts are indispensable; they cannot be replaced by other concepts since their use for the description of our experience is the precondition of that description being objective and our communication being unambiguous. Bohr's grounds for holding this view will be pursued further below.

Bohr's reply to Born and what seems to be his acceptance of Born's suggestion of the structure of mathematical invariance as constituting the reality behind the phenomena may support another interpretation of Bohr's concept of reality. One of Bohr's later students and assistants, Aage Petersen, has argued that the creation of quantum mechanics can be seen as an attempt to abandon every form of ontological thinking. His view may properly be called ontological nihilism. For in quantum physics there is no question of investigating an ultimate reality but of determining a precise use of a formal language. In fact he claims that it is feasible to say that quantum physics is close to being a part of mathematics, and that relinquishing ontological claims in quantum mechanics corresponds to abandoning ontology in mathematics, as was the case in classical antiquity. He also claims that the introduction of the principle of correspondence in particular involves a revolt against the ontological roots of earlier philosophies. The principle of correspondence embodies the notion of a mathematical generalization, and the use of this concept in the invention of quantum mechanics creates the formal similarity between the rise of quantum physics and the development of mathematics by the Pythagoreans.[42]

The concept of generalization as a part of the principle of correspondence must be understood, Petersen says, as an extension of the application of a set of concepts so that the concepts can be used in other circumstances. He seems to believe that if it is possible to generalize a system of concepts along these lines, one may also liberate the system from its ontological implications and preconditions. This was in fact what happened in quantum mechanics when the classical framework was generalized to include atomic phenomena as well as in mathematics where the concept of number was extended so as to contain irrational numbers. Given such a generalization all ontological stipulations are irrelevant since what is consequently in focus is the achievement of an unambiguous and exhaustive description.[43] Indeed, Petersen maintains that there exists a requirement that a physical theory has to meet that goes beyond the demand for consistency: since physics is an experimental science, the predictions of a theory have to be in harmony with the observational facts. But these requirements are the only two whose satisfaction is mandatory in the creation of physical theories.

What is aimed at in quantum mechanics is not intuitive understanding of the physical content of the theory but the unambiguous use of the relevant concepts. The physical content, Petersen says, may be completely non-visualizable and inaccessible to intuitive understanding and therefore the theory need not conform to any ontological commitment.[44] The most characteristic feature of the quantum mechanical formalism is the non-commutativity of canonically conjugate variables, which implies that a pair of such variables cannot be simultaneously well-defined. The existence of this property also explains why quantum mechanics cannot be "ontologized" while classical mechanics apparently could be. The explanation is that in classical physics, contrary to what is the case in quantum physics, all dynamical and kinematical variables commute, so there is no problem concerning what meaning might be attached to undefined or merely partially defined variables.[45] In quantum mechanics, however, the course of an electron entering an interferometer is not well-defined in the interval spanning its source and the point at which it impinges on the photographic plate because at least one of the disjuncts, the momentum or the position, is ill-defined. Hence it is senseless to propose an ontological interpretation of quantum mechanical statements containing such ill-defined expressions. As Petersen says, "What mode of being, if any, can be ascribed to something that is partly undefinable?"[46]

Furthermore, Aage Petersen claims that the theory of relativity has already proved to be a serious challenge to the ontological interpretation of mechanics. The theory implies that the size and shape of an object or the space-time interval between events have no absolute significance but that these properties depend on the velocity of the system of reference in which they are observed. "These concepts are not numerically well defined until the observer has specified his state of motion relative to the system".[47]

Petersen's conclusion is then that quantum mechanics is in conflict with an ontological way of thinking. The concept of reality can only be specified in relation to a conceptual framework, and it is the logical structure of this framework which is central to the discussion. What we are looking for in quantum mechanics is an unambiguous description of atomic phenomena under certain experimental conditions, not an ontological characterization of reality. So quantum mechanics has provided us with an insight into the relation between language and reality. It shows us that the language-reality problem is fixed by deep-rooted logical features of the conceptual framework itself; and an analysis of these features reveals the concept of reality to be related to the elements of arbitrariness in the physical description. This is to say that the elements which are not part of the algorithm, the various experimental preparations of the initial conditions and the succeeding collapse of the wave function, seem, at least for the time being, "to require conceptual reference to something 'outside' the algorithm".[48]

Ontological nihilism is definitely not part of Bohr's thought, and, needless to say, Petersen's arguments for a withdrawal of ontological commitments from quantum theory are deeply confused and so fail. He argues that since dynamical terms in quantum mechanics do not have the same absolute meaning as they had in classical physics, their referents cannot be attributes possessed by the things in themselves. Consequently we have to abandon the ontology of classical physics. But this conclusion does not mean that we have to abandon ontology altogether. This would be an extraordinary *non sequitur*. A claim that a certain physical magnitude is relative is equivalent to saying that an object has this property when it is related to other objects in the appropriate circumstances; or, in other words, that the attribution of such a magnitude to the object is context-dependent. But this is not the same as saying that it is not real or is subjective. The magnitude is real in the sense that the atomic object manifests itself as having this property in relation to a certain experimental arrangement. That kinematical and dynamical properties are ill-defined in situations where a couple of measurements have not been made does not differ from the meaningless use of any other concept, all of which have to conform to certain conditions in order to be used unambiguously. And in those cases in which such a concept stands for a relative property, like the position concept and the momentum concept, it means that the conditions for its use are context-dependent instead of being context-independent.

Petersen is right insofar as he argues that the quantum mechanical formalism which includes expressions of operators, wave functions, superposition and so on does not give us a description of the physical world as it is. This is also what Bohr holds. The entire complex of general theoretical sentences which makes up the core of quantum theory cannot be assigned any truth-value whatsoever. The theory specifies linguistic instructions for the description of certain areas of our experience. But Bohr likewise maintains that experimental statements of quantum mechanics have, in appropriate circumstances, a determinate truth-value, and hence describe the real world. And this is because ascribing a truth-value to a belief or a statement on the basis of the recognition that the appropriate circumstances obtain is the same as asserting to be real those states of affairs which are judged to confer that value. Hence Petersen is mistaken in believing that quantum mechanics does not entail ontological commitments. Even for an anti-realist, material-object statements which are used to report experimental observation involve a certain number of commitments. This is what Bohr has in mind when he says: "The renunciation of pictorial representation involves only the state of atomic objects, while the foundation of the description of the experimental conditions, as well as our freedom to choose them, is fully retained".[49]

It might be objected, nevertheless, that this is correct only as long as truth conditions and meaning conditions are identified. Bohr's references are to the conditions for unambiguous communication, a question of meaning, not of truth. This is evident, but since Bohr was a physicist, not a philosopher, he was concerned with the conditions for an unambiguous use of classical descriptive concepts in the domain of quantum mechanics and was not concerned with how people generally grasp the meaning of descriptive sentences. This may explain why he did not speak of truth *simpliciter*; he never considered what might qualify as a reasonable candidate for a theory of the speaker's understanding of a linguistic expression. What he did formulate was a semantic theory of how the meaning of quantum mechanical sentences is determined by relating the correct use of such expressions to certain recognizable conditions. However, these conditions may well be called the immanent truth conditions that hold for such a statement where the notion of a truth-value is accounted for in terms of the assertibility of a statement.

The ontology to which Bohr is committed on the basis of his semantic theory is indeed that of anti-realism, but the objective version of it. The semantic argument to which Bohr subscribes is that the atomic object can be meaningfully ascribed a determinate dynamical or kinematical property only if some truth-conferring evidence is available for assessing such predicative statements. But the grounds available for the assertion of a particular statement are not exhaustively connected to what can be immediately experienced. In general, Bohr holds that the objects of our perception, the phenomena, are not objective or well-defined in themselves; it is first when, and only when, they are grasped in virtue of their subsumption under the concept of continuity that the content of our experience can be said to be a concern with the real. The unambiguous communication of a particular physical experience is then at hand if we express what we observe in a language which makes provision for a description of a causal spatio-temporal connection between the content of this experience and the content of other possible experiences. Bohr also denies, however, that the meaning of statements that are used to communicate our physical experience is determined by the theoretical or conceptual framework alone, this being one of the basic assumptions of a radical ontological relativism as well as ontological nihilism. The conditions for the meaningful application of classical concepts which are expressed by these statements are conditions which can be observed. As noted above, the general statements of a scientific theory are linguistic rules for descriptions which define the possibilities of unambiguous communication, but in any particular case the conditions for the proper use of these rules are indeed empirical. It is experience which determines which of the possible descriptions is the correct one to use in the actual case by judging which of the conditions that endow a given utterance with a certain determinate truth-value are realized. But it is not only experience, because the same experience may be described in different ways depending on the use to which the description is put. I think this is what Bohr meant by saying that the subject-object distinction was a movable partition. It is pragmatic factors – not ones in nature – which determine where the micro-macro cut is made and thereby determine which parts of the experimental arrangement the reference to which counts as the conditions for a meaningful predication and which parts belong to the atomic object.

So, although Bohr never defined truth and rarely used the term, an empirical as well as a pragmatic component was built into his notion of truth, as was the case with Høffding's conception. A particular descriptive statement in physics predicating a certain attribute of an object is claimed to be true if, and only if, (1) the satisfaction of the experimental conditions for asserting that sentence are, or can be, empirically confirmed, and (2) the sentence coheres with other individual descriptive sentences in virtue of the fact that they each follow the rules for description, which are fixed by a consistent theory incorporating classical concepts. Since only material-object statements are capable of serving as reports of observation that are thus decidable, only such statements can express what is objectively real.

Let us look more closely at Bohr's conception of the assertibility conditions and at the reasons why he does not subscribe to the notion of verifiability as constituting the sole warrant for assertibility. Bohr says that we are epistemically justified in applying the classical concepts in the description of phenomena in the situations where such concepts refer to what can be observed. Since exact position phenomena physically exclude exact momentum phenomena, the classical concept of momentum refers to nothing in describing a position phenomenum. But such phenomena are certainly regarded as being related to preceding and subsequent measurements (preparation and detection), and so, consequently, the form that that connection takes, viz. the quantum system or atomic object, has reality. So even when unobserved there is a quantum mechanical system which has reality but which cannot be described as possessing the classical state-defining properties. When observed in interaction it can be described in either kinematic or dynamic terms, but these descriptive concepts refer to objective relations, yielded by the whole experimental set-up, including some actual measurements. Thus we have two types of experimental statement in quantum mechanics: one comprising those which are concerned with a synchronic description from subject to object at each instant by predicating a certain attribute to an atomic object, and a second comprising those concerned with a diachronic description from an observational state of measuring at one instant to a second state at any other instant by ascribing these states to one and the same object.

So Bohr's semantic theory contains two types of explanation of how the various elements of an experimental sentence contribute to its meaning. The meaning of the experimental statements is partly constituted by their verifiability conditions, which enable us to discover their truth-value. This is because the meaning of kinematic and dynamic predicates consists solely of those truth conditions of whose satisfaction we in fact have knowledge or the ability to acquire it. But, contrary to the experimental statements of classical mechanics, which are true if they can be verified, the experimental statements of quantum mechanics, of the kinds under discussion, are true only if they are actually verified. Thus Bohr would say that statements ascribing dynamical or kinematical properties to an atomic object have investigation-dependent truth-values contrary to the investigation-independent truth-values of the similar statements of classical mechanics. This is what he means by claiming:

> The emphasis on permanent recordings under well-defined experimental conditions as the basis for a consistent interpretation of the quantal formalism corresponds to the presupposition, implicit in the classical account, that every step of the causal sequence of events in principle allows verification.[50]

Obviously, the conditions for the correct application of classical state concepts change from classical mechanics to quantum mechanics in the sense that in quantum mechanics the satisfaction of these conditions consists of some *actual* verification, owing to the fact that the course of the atomic object in space and time cannot be subdivided, while in classical mechanics the satisfaction is connected to some *possible* verification, since here such a subdivision can always, in principle, be empirically confirmed and thereby form the basis on which the definition of a classical state of an unobserved particle is justifiable. Thus, owing to the quantum of action and the uncontrollable interaction between the atomic object and measuring instrument there is, contrary to classical mechanics, no way in which we may possibly be able to confirm a putative definition of a classical state and hence nothing which can warrant the ascription of such a state to the *unobserved* atomic object. The predication of a state-defining property of the *observed* atomic object is what *is* verifiable, and consequently, the assessment of the value of one of the classical state-defining parameters precludes the assessment of the other as meaningful.

However, the concept of an atomic object figures in both types of the experimental statements of quantum mechanics. A particular kinematic or dynamic value is ascribed to an atomic object under such conditions just mentioned, but the meaning of the term "atomic object" is not related to the satisfication of the actual conditions of verification. To understand the term "atomic object" in a given experimental statement is indeed to know how the occurrence of this term in the sentence contributes to the determinination of the truth condition of the sentence. In responding to this Bohr would assert that the correct use of this theoretical concept is determined by the entire quantum theory which, if we follow the descriptive rules being expressed by it, allows us to describe a number of experiences and to regard them as manifestations of the atomic object. That is to say, the concept of atomic object is defined by all the possible relations which connect it to the concepts of observables within quantum theory. Thus part of the meaning of an experimental statement, in which a particular property is predicated of an atomic object as observed or in which the result of one measurement is connected with the result of a subsequent measurement, *qua* being described as recordings of one and the same object, consists in its conditions of derivability from the formalism. This means that experimental statements attributing properties to the atomic object have to be deducible from the quantum theory if they are to be counted as having a truth-value. We may understand the meaning of such experimental statements only if we are able to point to a procedure in which they can be derived from a consistent theory. So the correct use of experimental sentences is in part determined by a theory for which we have strong grounds for its empirical adequacy in virtue of its predictive force.

I believe therefore that Bohr looked at the reality of the atomic object in the following way: what we observe is the atomic object in a certain state; we do not observe a phenomenal object which is caused by the atomic object. Atomic objects do not cause their own states which we then perceive; they are in these states whenever we observe them. Such observation is accordingly interpreted on the assumption that the atomic object interacts with the measuring instrument in such a way as to make it the case that the state of the atomic object is made manifest by the state of the measuring apparatus. This latter state may be amplified by some devices with irreversible functioning so that the result of the amplification counts as a registration.[51] And it is only in virtue of this registration and knowledge of how the instrument works that we have sufficient grounds for holding that our ascribing a certain state to the atomic object is correct. But, in spite of the fact that the correctness of the ascription is determined by observation, the ascription is concerned with an objective state of affairs.

Consequently, Bohr holds the view that the atomic object sometimes possesses a particular, definite observable property, sometimes not, depending on whether given experimental measurements warrant our making the proper predication of it or not. This is so because the ascription of such observable properties is relative to an appropriate experimental arrangement, that is to say it is context-dependent, although Bohr, furthermore, believes that the ascription is also relative to certain actual recordings. The significance of calling a property relative with respect to external circumstances is that of pointing out that it can only be predicated of the object if certain experimental conditions are fulfilled. A relational property is not a quality that has been possessed prior to the object's entry into that relation. This claim too implies that not all relational, although observable, properties can be ascribed to an object at once. Certain recognizable conditions have to be satisfied before this can be done. Bohr maintains, indeed, that a relational property of the atomic object has to be observed in order to be predicated of it. The impossibility of the simultaneous ascription of all relational properties does not, however, confer any support to the assumption that Bohr held that the object, which is described in terms of quantum mechanical statements, is, in consequence, the phenomenal object. On the contrary, the phenomenal object cannot exemplify relational properties since the phenomenal object is merely its actual appearance. One and the same phenomenal object cannot appear differently in different experimental circumstances, only the atomic object can. But neither does this mean that Bohr went to the opposite extreme, claiming that the atomic object is something behind its various phenomenal appearances. The atomic object does have several possible phenomenal manifestations in the sense that it can only meaningfully be ascribed several relational properties with respect to certain experimental conditions which mutually exclude each other. But this does not entail that the nature of the atomic object is something over and above its various possible manifestations. A phenomenon, in Bohr's terminology, is not an object but the property with which it becomes visible with respect to a certain experimental situation. So the phenomenal object is the atomic object exhibiting a particular property. From this it does not follow that the atomic object is more than its various observable properties, or that it is more than what the various complementary descriptions may tell us, or indeed, that Bohr believed it was. He did not.

For these reasons I do not believe Murdoch is right in claiming that Bohr's theory of measurement is based on the assumption that it is pre-existing values which are being recorded by the position experiment or the momentum experiment. Bohr definitely maintains that there exists a correlation between the value of the atomic object and that of the measuring instrument: "A measurement can mean nothing else than the unambiguous comparison of some property of the object under investigations with a corresponding property of another system, serving as a measuring system".[52] This is also the passage on which Murdoch bases his construction "according to which successful observation or measurements reveals the objective, pre-existing value of an observable".[53] But, if the registered value is the value of the state of the atomic object immediately prior to measurement, then the atomic object must either be in the state whose value is observed because it is an inherent state, or because it is an objective relational state which is then supervenient on some inherent state of the atomic object as well as on the state of the measuring instrument. Such an assumption has no correlate in Bohr's thinking. It might have been part of his thinking if he had made a distinction between the atomic object as thing-in-itself and as thing-as-it-appears and if he had continued to subscribe to the meaningfulness of the notion of measurement disturbance. And in fact Murdoch believes Bohr did intend this.[54] But as an interpretation of Bohr's ontology it cannot be correct, as I have already argued. For even though Murdoch makes the observation that Bohr even after 1935 spoke of the interaction between the atomic object and the measuring instrument, he has not proved that Bohr understood, by this expression, an entity having such independent substance possessing determinate properties or having an inherent state so as to warrant the strongly realistic interpretation he gives it: in order for a measurement to be made the object has to act upon the instrument, which then reacts in turn, and consequently disturbs the object. In support of his construal Murdoch quotes two passages from Bohr's reply to Einstein, Podolsky and Rosen in which he speaks about "the impossibility of controlling the reaction of the object on the measuring instruments". But such evidence carries little weight since at that time Bohr was at a watershed, just about to revise some of his main arguments for complementarity.

There is no doubt that Bohr regarded the measured value of an observed property as an objective one in the sense of being mind-independent. Heisenberg's subjectivist suggestion, according to which the observer himself takes part in the determination of the state through his reading of the measuring instrument, is entirely dismissed by Bohr. For him "it is certainly not possible for the observer to influence the events which may appear under the conditions he has arranged".[55] Nor does the measuring instrument by itself create the observed value. But this does not entail that Bohr assumed that what is measured is a pre-existing value or a disturbed value. As we saw when discussing Bohr's reaction to the EPR thought experiment, he argued in his Warsaw lecture that the concept of a state is ambiguous as long as it is used without specific reference to the experimental conditions. Therefore in order to avoid inconsistencies in his interpretation of Bohr's theory of measurement, Murdoch has to claim that "observables which are not measured cannot meaningfully be said to have pre-existing values".[56] This claim is nevertheless peculiar. How is it meaningful to hold, at the same time, that the value of an observable to be measured is the pre-existing value of the property of the object which exists *prior to* the measurement and likewise to hold that this value does not exist if it is not measured? Murdoch seems here to be forced to maintain, whether he likes it or not, that the observable to be measured is supervenient on some pre-existing state of the atomic object as well as of the measuring instrument in order for the measured value to be a value an object possesses immediately before it is measured. But such an assertion is not consistent with Bohr's idea that none of an EPR pair of objects can be ascribed a certain definite property until its value is directly measured at one of them. It is my belief that in Bohr's view the measured value of a certain property is not a pre-existing value, as Murdoch claims, nor is it a produced value obtained through disturbance caused by the measuring instrument, but it is one which arises in the interaction between the atomic object and an appropriate measuring instrument.

But, one may ask, how is it possible for Bohr to think of the ascription of a dynamical or kinematical variable as objective and at the same time hold that such a predication merely possesses an investigation-dependent truth-value? If it is not so that an atomic object has any state-defining property prior to our investigation, it is obvious that a statement attributing a certain value to the object cannot be correctly assigned a determinate truth-value unless we actually discover whether a measurement makes it true or false. However, the investigation-dependent truth-value of state-defining statements in quantum mechanics can hardly disqualify them from expressing objective states of affairs. For each time we predicate a certain observable property to an atomic object there exists a definite reproducible procedure which we must follow in order to make a meaningful ascription and which allows us to determine the truth-value of the predicative statement. This is due to the fact that the conditions for a correct application of a dynamical or kinematical concept in quantum mechanics is limited to an appropriate experimental investigation which makes these concepts stand for relational properties.

Any statement that predicates a relational property to an object may be given either a realistic or an anti-realistic construal. On the realist view a relational-property statement expressing the fact that *a* stands in relation *R* to *b* is true if *a* and *b* possess some inherent properties independently of each other on which *R* supervenes and which may, indeed, be undiscoverable. The anti-realist has to say that such properties on which the relational properties supervene must be possessed independently of each other and so they must, in principle, be accessible to our cognitive powers in order for us to ascribe any investigation-independent truth-value to a statement predicating such properties to an object. But Bohr subscribed to the strong meaning condition by not merely restricting the correct use of a kinematical or dynamical observable to the presence of the appropriate measuring instrument. So he has to argue that since an atomic object does not possess these observables as essential or inherent properties, but exemplifies such properties only in relation to particular acts of measurement, then these relational properties obtaining in virtue of the interaction between the object and the measuring instrument cannot supervene on non-relational facts of the relata, whether they are observable or otherwise. In these circumstances the atomic object is a totality whose existence is knowable merely through its non-derivative relation properties. This means, of course, that the property which is given in the interaction between the atomic object and the measuring instrument cannot supervene on non-relational properties of the cause and the effect, that is to say, on the atomic object as the cause and the recordings of the measuring instrument as the effect. Instead it is supervenient on the entire experimental set-up, including certain measurements.

Thus when Bohr questions "the reality of the atomic object in the ordinary physical sense" he questions it as a material corpuscular localized in space and time and maintaining deterministic causal connections with its previous and its subsequent states. His question concerns what may count as facts about the atomic object. He did not question its reality *tout court*. Bohr thought of atomic objects as real mind-independent entities to which the theoretical term "atomic object" in experimental statements refers. So when he said that quantum mechanics throws "new light upon the old philosophical problem of the objective existence of phenomena independently of our observations" it was in order to revise our understanding of the objective existence of independent entities, not to call their existence into question. Nevertheless, he was not a realist because he denied that quantum theory could in principle provide us with any true assertions of this atomic object other than what can be obtained on the basis of direct observation. The meaningful application of a certain description to an atomic object is determined by the outcome of the experimental measurements by yielding us cognitive grounds for the predication of a certain property of it. It is the result of two distinct measurements described according to the formal rules of quantum theory which confers determinate truth-values on experimental statements of quantum mechanics. Bohr's belief was, in sum, that atomic objects are real but their mode of existence is dependent on our cognitive faculty.

Epilogue: The Legacy

Bohr was an objective anti-realist, in contradistinction to other contemporary anti-realists. As a student he had found his way to this position through the philosophical training given him by Høffding. Thus anti-realism was his heritage. The young Bohr shared with Høffding an adherence to the same criterion of reality derived from the belief that perceptual experience could be accounted for by a conceptual framework which describes the phenomena as entering into causal connections in space and time and is as such concerned with what is real; they both regarded the resulting subject-object distinction as fundamental to an analysis of the conditions for scientific knowledge; they both thought that in areas where the subject interacts with the object, making the criterion of reality inapplicable, what they called an "irrational element" will manifest itself to our cognition; they both thought of this irrational element as something which could be handled by using "complementary" modes of description, for Høffding in the fields of psychology and ethics, for Bohr in that of quantum mechanics. These various elements were parts of the anti-realistic message Bohr received from his mentor. And it is testimony to Bohr's greatness that he was able to transform what had been passed on to him by giving it a firm foundation upon which his interpretation of quantum mechanics might be based, a foundation so solid that this interpretation has been far superior to any other attempt, realist or anti-realist, that has been made until now.

Most philosophers or physicists who have examined Bohr's philosophy have identified it with some form of instrumentalism, phenomenalism, positivism or Kantianism. A minority of philosophers, especially in recent years, have defended a realistic interpretation of Bohr's view on quantum mechanics. That there are so many different construals of complementarity is not so very astonishing considering the roots of complementarity. As the great eclectic philosopher he was, Høffding was possessed of penetrating insight with respect to both the rationalism and the idealism of the German tradition, the positivism of the French tradition, the empiricism of the British tradition and the pragmatism of the American tradition, and he combined what he in each thought contributed to the best and ultimately most satisfying way of addressing the problems generated by the various philosophies. If his combination of rationalism and empiricism was to be both coherent and successful Høffding had to sacrifice some features of the different schools, such as the idea of forms of thought being *a priori* categories in the Kantian sense and the idea of a reality existing behind experience to which true descriptive statements refer. Therefore, his notion of truth became a multi-facetted concept, splicing together both empirical and pragmatic elements. Since Bohr's viewpoint of complementarity has its source in Høffding's eclectic philosophy, it is little wonder that different philosophers have characterized Bohr's philosophy in so many different ways. For these philosophers have been confronted with a philosophy which had its roots in a position that did not fit into a single traditional philosophical classification, but yet shared features which were thought to belong to diverse schools. It is therefore easily understandable why most of his interpreters have been confused by this and have stressed one or the other of what they saw as positivistic, idealistic or phenomenalistic features of complementarity owing to Bohr's rejection of the theory of truth as correspondence and his denial of a substance-property ontology, or why some have been misled by his statements about the reality of atoms, maintaining a realistic interpretation of complementarity.

There is, however, also another legacy left by Bohr, one which cannot so easily be traced back to Høffding. This legacy takes the form almost of a credo in one of his essays:

> As the goal of science is to augment and order the experience, every analysis of the conditions of human knowledge must rest on considerations of the character and scope of our means of communication. Our basis is, of course, the language developed for orientation in our surroundings and for the organization of human communities. However, the increase of experience has repeatedly raised questions as to the sufficiency of the concepts and ideas incorporated in daily language. Because of the relative simplicity of physical problems, they are especially suited to investigate the use of our means of communication. Indeed, the development of atomic physics has taught us how, without leaving common language, it is possible to create a framework sufficiently wide for an exhaustive description of new experience.[1]

As a result of his reflection on quantum mechanics and on the use of classical concepts in the description of atomic objects, Bohr at length concluded that any intelligible epistemology has to be based on a semantic theory. He acknowledged that the limit of our knowledge of atomic objects, owing to the quantum of action, is entailed by the change of conditions required for a meaningful application of classical descriptive concepts. But his semantic theory incorporated, of course, the earlier anti-realist elements which are to be found in his epistemology.

Thus Bohr's anti-realism in its most fully developed version is based partly on a general argument supportive of semantic anti-realism, to the effect that in the domain of quantum physics we can communicate meaningfully only about that of which we are able to obtain empirically based and assured knowledge. The sense of classical concepts, which are merely refinements of those concepts that are entrenched in our thought and expressed in ordinary material-object language, does not have a scope over and beyond what our cognitive powers may guarantee. The use of classical concepts is meaningful only to the limits of what our cognitive powers can establish as certain. And any attempt to extend their use beyond these limits produces ambiguous descriptions, for in such a case there would be no precise and recognizable conditions for their application. For Bohr the conditions which determine the meaning of classical concepts in the domain of the quantum of action are related to the experimental arrangement, and only if a set of measurements are actually carried out do we have sufficient cognitive warrant for ascribing a certain property to the atomic object involved. This does not mean that the atomic object is not real. It is real in the sense that the theoretical term 'atomic object' in a statement which predicates a particular dynamical property to an atomic object refers to an objective reality. But that statement tells us nothing about an object in itself behind the manifestations which we directly observe. The observation is about something real because it can be lawfully connected with other observations through being described according to the rules of quantum theory.

Everyday language with its vocabulary of material objects is, on Bohr's view, the language in which we are able to form unambiguous statements about our ordinary experience. The rules and categories which are entrenched in this language, in that they make up the conceptual framework within which it operates, do in part constitute our ordinary experience by determining the forms under which we can speak meaningfully about our experience. Among those concepts that are part of the structure of our thoughts and are expressed in everyday language are time, space and causation, and in virtue of being so they acquire a sort of *a priori* status - which may be called pragmatic *a priori* - similar to that Høffding had assigned them. The rules and categories of ordinary language are necessary for us to communicate unambiguously about the world in which we are, in Bohr's terms, both actors and spectators. Only by means of this language are we able to describe to ourselves and to our fellow creatures what we experience and how we interact with the environment. All practical and experiential knowledge is expressed in ordinary language, and it can be expressed only in that language, because our knowledge is partly a result of it. We have therefore to adhere to such forms of expressions when we attempt to describe new areas of experience which have not yet been described. Science has led to mathematics in order to attain to a precise description of our experience. But even mathematics is a refinement of ordinary language, although operating on a level of abstraction which may have no direct reference to our experience. Mathematics is a useful addition to it, "supplementing it with appropriate tools to represent relations for which ordinary verbal expression is imprecise and cumbersome". So in discovering new areas of experience we may find that the content of some of the categories and rules entrenched in natural language have to be modified if their use in the light of the novel experience proves to be ambiguous and inaccurate. In cases where what has hitherto constituted the conditions whose satisfaction warranted assertions descriptive of our observations fail to guarantee such coherent descriptions, on account of the fact that the epistemic grounds for their justification are lacking, the specification of a fresh set of conditions is required whose satisfaction restores coherence and thereby unambiguous communication. We have seen this to be the case in quantum mechanics: in order for it to be possible to speak meaningfully about atomic objects, some of our most elementary concepts have to undergo revision with respect to features deriving from their conditionality upon what was thought to hold in classical physics. This discovery is the true legacy of Niels Bohr, and one on which it is mandatory for every philosophy of science to take a position.

Notes

PROLOGUE

1. A photograph showing Bohr's desk and the paintings is published in N. Blædel, *Harmoni og Enhed. Niels Bohr, en biografi.* (Harmony and Unity. Niels Bohr, a biography). Copenhagen, Rhodos 1985, p. 152.
2. See, for instance, S. Rozental (ed.), *Niels Bohr: His life and work as seen by his friends and colleagues*, p. 13. North Holland Publishing Company. Amsterdam 1967; D. Favrholdt, "Niels Bohr and Danish Philosophy", *Danish Yearbook of Philosophy*, 13, 1976, 206–220; and D. Favrholdt, "On Høffding and Bohr. A reply to Jan Faye", *Danish Yearbook of Philosophy*, 16, 1979, 73–77.
3. *Archive for the History of Quantum Physics.* Interview with Professor Niels Bohr on 17th November 1962. The circulating transcript of the last interview is in some places inaccurate and deficient compared with the records on the tapes.
4. H. Folse, *The Philosophy of Niels Bohr*, North-Holland, Amsterdam 1985, p. 51 f.
5. See Aa. Petersen, "The Philosophy of Niels Bohr", *Bulletin of the Atomic Scientist*, xix, No 7, September 1963, 8–14, p. 14; and L. Rosenfeld, "Niels Bohr in the Thirties", in S. Rozental (ed.), *Niels Bohr*, 114–36, p. 124.

CHAPTER I

1. See P. Lübcke, "F.C. Sibbern: Epistemology as Ontology", in *Danish Yearbook of Philosophy*, 13, 1976, 84–104.
2. See S.E. Stybe, "Niels Treschow (1751–1833), A Danish Neoplatonist", in *Danish Yearbook of Philosophy*, 13, 1976, 29–47.
3. H. Høffding, *Erindringer* (Memoirs), Copenhagen 1928, p. 51. All the translations from Danish and French are those of the present author.
4. *Ibid.*, p. 62.
5. H. Høffding, "En filosofisk Bekendelse" (A Philosophical Confession), p. 24, *Mindre Arbejder III*, 18–27.
6. H. Høffding, *Erindinger*, p. 171 f.
7. N. Bohr, "Mindeord over Harald Høffding" (Harald Høffding in Memoriam), *Oversigt over Det kgl. Danske Videnskabernes Selskabs Forhandlinger 1931–1932*, 131–36, p. 131.
8. N. Bohr, "Physical Science and the Problem of Life", in *Atomic Physics and Human Knowledge* [abbr. *APHK*], New York 1958, 94–101, p. 96.
9. H. Høffding, "Mindetale over Christian Bohr" (Commemorative speech on Christian

Bohr), *Tilskueren* 1911, 209–12.
10. See Rasmus Nielsen, *Almindelig Videnskabslære i Grundtræk*, Copenhagen 1880, pp. 172–83.

CHAPTER II

1. P. Skov, *Aarenes Høst. Erindringer fra mange Lande i urolige Tider* (Harvest of the years. Recollections from many countries in times of unrest), Copenhagen 1961, p. 10.
2. V. Slomann, "Minder om samvær med Niels Bohr" (Recollections of Niels Bohr). A feature article in the newspaper *Politiken*, 7.10.1955.
3. See S. Rozental (ed.), *Niels Bohr*, p. 24 f.
4. See *Aarbog for Københavns Universitet 1904-07* (The Yearbook of the University of Copenhagen 1904-07), Copenhagen 1911, p. 122.
5. Slomann goes on to say: "But in the autumn of 1920 some of its former members met up as guests on a farm in South Zealand." The farm must have been "Grubberholm", the home of Astrid Lund and Elias Lunding.
6. See J. Witt-Hansen's investigations concerning the Ekliptika Circle in his "Leibniz, Høffding, and the "Ekliptika" Circle", *Danish Yearbook of Philosophy*, 17, 1980, 31–58.
7. Høffding's trip to America is thoroughly described in his Erindringer, p. 202–14.
8. See Den filosofiske eksamen 1849–1911. Hovedeksamensprotokoller in the National Archive. K.U.35.18.01-05.
9. Included among the papers, now in the Royal Library, belonging to Harald Høffding's son, Hans Høffding, is a handwritten inventory of his lectures and seminars from 1871 to 1912. The other source is a printed list in the Yearbooks of the University. There is a slight but most significant difference between these two sources for the spring of 1905, the term in which *Ekliptika* most certainly was founded. The hand-written inventory only mentions two series of lectures in this term: 1) The psychology of free will; 2) Philosophical theories (on the basis of modern philosophers); but in *Aarbog for Københavns Universitet 1904-07*, on page 122, we read concerning the spring term of 1905 that Høffding provided three series of lectures: 1) The psychology of will; 2) Philosophical theories in recent times; 3) Lectures on Kierkegaard. Since the yearbook also mentions the attendance figures for each series, it must, on this point, be more reliable than the handwritten inventory.
10. H. Høffding, "Begrebet Villie" ("The concept of will"), *Mindre Arbejder III*, Copenhagen 1913, 28–52.
11. "Last interview", Niels Bohr Archive, transcript, p. 1.
12. G. Cohn, "Harald Høffding og hans Filosofi" (Harald Høffding and his Philosophy), *Tilskueren*, 50, 1933, 103–17.
13. W. James, *The Principle of Psychology I-II*, London 1891, Vol I, p. 206.
14. G. Holton, "The Roots of Complementarity", *Daedalus*, 99, 1970, 1015–55, p. 1035.
15. D. Favrholdt, "Niels Bohr and Danish Philosophy", pp. 217–18.
16. Last interview, Niels Bohr Archive, transcript, p. 1 f.
17. *Archive for the History of Quantum Physics*. Interview with Oscar Klein on 20th February 1963, number 2 of 6 sessions, conducted by Th.S. Kuhn and J.L. Heilbron; quoted from p. 9 of the transcript.
18. Letter of 20 April 1909 from Niels to Harald, published in Niels Bohr, *Collected Works*, Vol. 1, p. 501, North-Holland Publishing Company, Amsterdam 1972-.
19. Letter of 26 April 1909 from Niels to Harald, published in *Collected Works*, Vol. 1, p. 503.
20. J. Rud Nielsen, "Memories of Niels Bohr", *Physics Today*, 16, 1963.

21. See M. Jammer: *The Conceptual Development of Quantum Mechanics*, p. 178 ff, McGraw-Hill, New York 1966; and G. Holton, "The Root of Complementarity".
22. Harald Høffding, *Søren Kierkegaard som Filosof* (Søren Kierkegaard as philosopher), Copenhagen 1892, p. 74. Quoted from the 2nd edition 1919.
23. H. Høffding, *A History of Modern Philosophy*, Dover, New York 1955, II, pp. 287–88.
24. *Ibid.*, p. 79.
25. Holton, "The Roots of Complementarity", p. 1042.
26. H. Høffding, *Den menneskelige Tanke, dens Former og dens Opgave* (Human Thought, Its Forms and Its Tasks), p. 288, Copenhagen 1910.
27. *Ibid.*, p. 185.
28. See D. Favrholdt, "The Cultural Background of the Young Bohr", p. 452. *Rivistra di Storia della Scienza*, 2, no.3 1985, 445–61.
29. Niels Bohr Archive, *BSC*: 3.
30. Niels Bohr Archive, *BSC*: 3.
31. H. Høffding, "Det psykologiske Grundlag for logiske Domme", *Det kgl. Danske Vid. Selsk. Skrifter*, Sjette Række. Historisk og Filosofiske Afdeling, Fjerde Bind, Copenhagen 1893–99, 343–403.
32. H. Høffding, *Formel Logik* (Formal Logic), 4th edition, Copenhagen 1903.
33. Niels Bohr Archive, *BSC*: 3.
34. Niels Bohr Archive, *BSC*: 3.
35. H. Høffding, *Formel Logik* (Formal Logic), 5th edition, Copenhagen 1907.
36. Niels Bohr Archive, *BSC*: 3.

CHAPTER III

1. Niels Bohr Archive, *BPC*.
2. The letter is mailed from Bohr's institute at Blegdamsvej, which was built in 1920. Bohr mentions two names in it: Philipsen and Mayor Jensen. The former is probably Gustav Philipsen (1853–1925), who was alderman in the corporation of Copenhagen from 1909 until he died, and the latter must be Jens Jensen (1859–1928), who became one of the mayors in the corporation in 1903, a position he held until November 1924, when he became the Prefect.
3. F. Brandt *et al.* (eds.), *Correspondance entre Harald Høffding et Emile Meyerson*, Copenhagen 1939.
4. Niels Bohr Archive, *BPC*.
5. N. Bohr, "Ved Harald Høffding's 85-årsdag" (On the Occasion of the 85th Birthday of Harald Høffding), *Berlingske Tidende*, 10 March 1928. Evening edition of this newspaper.
6. E. Rubin, *Harald Høffding in Memoriam. Fire taler holdt på Københavns Universitet paa Harald Høffdings 89 Aars Dag 11. marts 1932*, (Four speeches made at the University of Copenhagen on Harald Høffding's 89th Birthday 11th March 1932).
7. N. Bohr, "Mindeord over Harald Høffding", p. 134.
8. See, for instance, N. Bohr, "The Quantum of Action and the Description of Nature" (1929), in *Atomic Theory and the Description of Nature* [abbr. *ATDN*], University Press, Cambridge 1961, p. 100; and "Biology and Atomic Physics" (1937), in *APHK*, p. 21–22.
9. N. Bohr, "Mindeord over Harald Høffding", p. 134–35.
10. See H. Høffding, "Bemærkninger om Erkendelsesteoriens nuværende stilling" (Notes on the Present State of the Theory of Knowledge), *Det kgl. Danske Vid. Selsk. Filosofiske Meddelelser II*, 2, Copenhagen 1930, pp. 9 ff. and p. 17.
11. Høffding's manuscripts, drafts and various notebooks are all kept at the Royal Library

in Copenhagen and run to many thousands of pages. The missing pages are from *Kladdebog XX* (Notebook XX), Ny kgl. Sml. 2053 fol., and they seem to have been written in the spring of 1928 just after Høffding had made a draft of his recommendation of Léon Brunschvicg for membership of the Royal Academy. This he must have done in January of 1928 because Brunschvicg was elected as a member on 13th April.

12. Niels Bohr Archive, *BSC*: 12.
13. This must be Bohr's Como lecture dubbed "The Quantum Postulate and the Recent Development of Atomic Theory" which was presented at Como in October of 1927 and which for the first time introduced Bohr's ideas of complementarity.
14. Niels Bohr Archive, *BSC*: 12.
15. See *Oversigt over Det kgl. Danske Videnskabernes Selskabs Forhandlinger 1927–28*, p. 26.
16. N. Bohr, "The Quantum Postulate and the Recent Development of Atomic Theory", in *Niels Bohr: Collected Works*, Vol. 6, p. 158
17. Niels Bohr Archive, *BSC*: 9. Published in *Collected Works*, Vol. 6, pp. 44–46.
18. Niels Bohr Archive, *BSC*:14. Published in *Collected Works*, Vol. 6, pp. 189–91 and pp. 430–32.
19. J. Kalckar, "General introduction to volume 6 and 7", in *Collected Works*, Vol. 6, p. xxvi
20. W. Heisenberg, "Quantum Theory and Its Interpretation", p. 107, in S. Rozental, *Niels Bohr*, 94–108.
21. Høffding's *Notebooks XX* and *XXI* contain the four drafts of this paper. I had claimed in my paper "The Bohr-Høffding Relationship Reconsidered", *Stud. Hist. Phil. Sci.*, Vol. 19, no 3, 321–46, note 39, that the various drafts were most likely written during 1929 or, maybe, during the winter 1929–30. I now think this is incorrect since the first draft is placed in *Notebook XX* not chronologically so remote from the missing paper, indicating that Høffding probably started on this essay not too long after his discussions with Bohr after the latter had returned from his summer cottage.
22. Niels Bohr Archive, *BSC*: 12.
23. See *Oversigt over Det kgl. Danske Videnskabernes Selskabs Forhandlinger 1929–30*.
24. See *The minutes of The Society for Philosophy and Psychology 1926–1963*, pp. 17–18, The Department of Philosophy, University of Copenhagen.
25. Published in *Collected Works*, Vol. 6, pp. 428–30.
26. Niels Bohr Archive: *Bohr's General Correspondence* (*BGS*).
27. Niels Bohr Archive: *BGC*.
28. H. Fuglsang-Damgaard, "Harald Høffding 1843–1943", *Dansk Teologisk Tidsskrift*, 6, 1943, 225–37, p. 237.
29. Niels Bohr Archive, *MSS*: 13. Some obvious misspellings by Bohr have been corrected.
30. Niels Bohr Archive: *BGC*.
31. "Bemærkninger om Erkendelsesteoriens nuværende Stilling", p. 17; parentheses and italics mine.
32. See *Den menneskelige Tanke*, p. 260.
33. See *Formel Logik*, p. 8 ff.
34. See H. Høffding, "Relation som Kategori" (Relation as a Category), p. 72, *Det kgl. Danske Vid. Selsk. Filosofiske Meddelelser I*, 3, Copenhagen 1921.
35. See *Ibid.*, p. 72.
36. See *Ibid.*, p. 75.
37. "Bemærkninger om Erkendelsesteoriens nuværende Stilling", pp. 18–19.
38. *Ibid.*, p. 19.
39. See *Ibid.*, p. 14 ff.
40. See *Ibid.*, p. 9 ff. and p. 17.
41. *Ibid.*, p. 28.
42. *Ibid.*, p. 19.

43. See *Ibid.*, p. 17.
44. See note 3.
45. *APHK*, pp. 116–17.

CHAPTER IV

1. *Erindringer*, p. 303.
2. H. Høffding, "Erkendelsesteori og Livsopfattelse" (The Theory of Knowledge and Apprehension of Life), p. 3. *Det kgl. Danske Vid. Selsk. Filosofiske Meddelelser II*, 1, Copenhagen 1925.
3. H. Høffding, *Filosofiske Problemer* (Philosophical Problems), p. 32. Copenhagen University Festskrift, Copenhagen 1902.
4. See *Ibid.*, p. 34.
5. *Ibid.*, p. 34.
6. See *Ibid.*, p. 35.
7. *Den menneskelige Tanke*, p. 136–37.
8. *Den nyere Filosofis Historie*, p. 49
9. *Ibid.*, p. 50.
10. See *Den menneskelige Tanke*, p. 132.
11. *Ibid.*, p. 258.
12. H. Høffding, "Begrebet Analogi og dets Filosofiske Betydning" (The concept of analogy and its philosophical importance), p. 33, *Mindre Arbejder II*, 33–46, Copenhagen 1904. This talk was published as "On analogy and its philosophical importance", in *Mind*, 14, 1905, 199–209. The quotation has been translated from the Danish edition.
13. See *Ibid.*, p. 103.
14. H. Høffding, *Psykologi i Omrids på Grundlag af Erfaringen* (An Outline of Psychology on the Basis of Experience), p. 280, Copenhagen 1882. Quoted from the sixth revised edition from 1911.
15. "En filosofisk Bekendelse", p. 25.
16. H. Høffding, *Moderne Filosofer* (Modern Philosophers), p. 84, Copenhagen 1904.
17. See *Psykologi*, p. 266.
18. See *Ibid.*, p. 269.
19. *Ibid.*, p. 268.
20. See *Ibid.*, p. 268.
21. "Relation som Kategori", p. 56.
22. See *Psykologi*, p. 268.
23. See *Ibid.*, p. 270.
24. *Den menneskelige Tanke*, p. 279.
25. See D. Favrholdt, "Bevidsthedsproblemet i Harald Høffding's filosofi" (The Problem of Consciousness in the Philosophy of Harald Høffding), *Det kgl. Danske Vid. Selsk. Historiske-filosofiske Meddelelser*, 44, 4, 1–36, Copenhagen 1969.
26. See *Psykologi*, p. 266 ff.
27. *Den menneskelige Tanke*, p. 209.
28. See *Ibid.*, p. 105.
29. See *Idem*; and H. Høffding, "Totalitet som Kategori" (Totality as Category), p. 48. *Det kgl. Danske Vid. Selsk. Skrifter*, 6. række, historisk-filosofisk afd. 2., Copenhagen 1917.
30. "Erkendelsesteori og Livsopfattelse", p. 94.
31. See "Bemærkninger om Erkendelsesteoriens nuværende Stilling", p. 3 f.; and *Den menneskelige Tanke*, p. 138 and p. 149.
32. See *Den menneskelige Tanke*, p. 257; and *Filosofiske Problemer*, p. 37

33. See "Erkendelsesteori og Livsopfattelse", p. 9; and "Totalitet som Kategori", p. 15.
34. *Den menneskelige Tanke*, p. 271.
35. "Totalitet som Kategori", p. 48.
36. See *Den mennekelige Tanke*, p. 234.
37. See *Ibid.*, p. 254.
38. See *Ibid.*, p. 258.
39. *Ibid.*, p. 261.
40. *Ibid.*, p. 260.
41. See *Psykologi*, p. 276.
42. "Relation som Kategori", p. 48.
43. See *Ibid.*, p. 45 ff.
44. See *Den menneskelige Tanke*, p. 274; and *Filosofiske Problemer*, p. 38.
45. H. Høffding, "Charles Darwin og Filosofien" (Charles Darwin and Philosophy), p. 217, *Mindre Arbejder III*, 202–228.
46. See *Den menneskelige Tanke*, p. 274 ff.; and H. Høffding: "Begrebet Analogi" (The Concept of Analogy), p. 89 ff., *Det kgl. Danske Vid. Selsk. Meddelelser I*, 4, Copenhagen 1923.
47. *Den menneskelige Tanke*, p. 278.
48. See *Ibid.*, p. 208 and p. 278 ff.; and *Filosofiske Problemer*, p. 43 ff.
49. *Den menneskelige tanke*, p. 281.
50. *Ibid.*, p. 209.
51. See *Ibid.*, p. 284 f.
52. *Filosofiske Problemer*, p. 48.
53. See *Ibid.*, p. 49; and *Den menneskelige Tanke*, p. 296.
54. See *Filosofiske Problemer*, p. 49.
55. See *Ibid.*, p. 50; cf. also "Relation som Kategori", p. 58 ff.
56. See *Den menneskelige Tanke*, p. 298.
57. *Ibid.*, p. 297.
58. See *Psykologi*, p. 84.
59. *Ibid.*, p. 278.
60. See *Filosofiske Problemer*, p. 51 ff.; and *Den menneskelige Tanke*, p. 299 ff.
61. See *Den menneskelige Tanke*, p. 301
62. F. Brandt, "Læreren og Humanisten" (The Teacher and the Humanist), p. 37, in *Harald Høffding in Memoriam*, pp. 31–39.
63. See *Den menneskelige Tanke*, p. 309.
64. See *Filosofiske Problemer*, p. 54.
65. See *Ibid.*, p. 55; and *Den menneskelige Tanke*, pp. 327–28.
66. See, for instance, J. Honnor, *The Description of Nature*, Clarendon Press, Oxford 1987, pp. 14–21 and pp. 170–75.
67. See *Den menneskelige Tanke*, p. 221 f.
68. "On analogy and its philosophical importance", p. 204.
69. *Psykologi*, p. 154.
70. *Ibid.*, p. 179 f.
71. *Den menneskelige Tanke*, p. 7.
72. *Etik* (Ethics), p. 122, Copenhagen 1887. Quoted from the third edition from 1905.
73. See *Den menneskelige Tanke*, p. 11 and p. 19.
74. *Psykologi*, p. 21; italics mine.
75. *Ibid.*, p. 29; italics Høffding's.
76. See H. Høffding, "Psykologi og Autobiografi" (Psychology and Autobiography), p. 14 and p. 20, *Det kgl. Danske Vid. Selsk. Filosofiske Meddelser II*, 3, Copenhagen 1943.
77. *Filosofiske Problemer*, p. 12.
78. See *Ibid.*, p. 17.
79. See *Den menneskelige Tanke*, p. 31.

80. See *Filosofiske Problemer*, p. 22.
81. See *Psykologi*, pp. 76 ff.
82. See *Ibid.*, p. 91.
83. *Ibid.*, p. 93; italics Høffding's.
84. *Den menneskelige Tanke*, p. 32.
85. *Psykologi*, p. 95.
86. See *Den menneskelige Tanke*, p. 31; cf. also "Begrebet Analogi", pp. 109–19.
87. See H. Høffding: "Spinozas Ethica", p. 377, *Det kgl. Danske Vid. Selsk. Skrifter*, 7. række, II, 3, Copenhagen 1918.
88. *Psykologi*, p. 287.
89. H. Høffding, "Om Vitalisme" (On Vitalism), pp. 48–9. Mindre Arbejder I, 40–50.
90. "Erkendelsesteori og Livsopfattelse", p. 23.
91. *Ibid.* p. 23. With respect to Høffding's criticism of vitalism, see "Totalitet som Kategori", paragraph 20.
92. *Ibid.*, pp. 26–27.
93. See *Moderne Filosofer*, pp. 81–98.
94. "On analogy and its philosophical importance", p. 201.
95. *Moderne Filosofer*, p. 83; and *Den menneskelige Tanke*, p. 183. The passage is from *The Scientific Papers*, Vol. I, Cambridge 1890, p. 156.
96. "On analogy and its philosophical importance", pp. 204–05.
97. Niels Bohr Archive, *BSC*: 3.
98. This paper had been published in the same year.
99. Niels Bohr Archive, *BSC*: 3.
100. "Begrebet Analogi", pp. 98–99.
101. Published in N. Bohr, *Collected Works*, Vol. 2, p. 584.

CHAPTER V

1. *ATDN*, p. 49.
2. N. Bohr, "On the Quantum Theory of Line-Spectra" (1918), *Det kgl. Danske Vid. Selsk. Skrifter. Naturvidenskabelig og Matematisk Afdeling*, Række 8, IV. Copenhagen 1918–1922, p. 8.
3. See *Collected Works*, vol. 3, p. 688.
4. N. Bohr, "On the Series Spectra of Elements" (1920), pp. 23–24, reprinted in *The Theory of Spectra and Atomic Constitution*, Cambridge University Press, 1922, 20–60.
5. *Collected Works*, vol. 3, p. 178.
6. See D. Murdoch, *Niels Bohr's philosophy of physics*, Cambridge University Press 1987, p. 39.
7. M. Jammer, *The Conceptual Development of Quantum Mechanics*, p. 116. The quotation is from N. Bohr, "On the Application of the Quantum Theory to Atomic Structure" (1922), *Proceedings of the Cambridge Philosophical Society*, Part I, Cambridge University Press, 1924, p. 22.
8. *Ibid.*, p. 42 ; parentheses and italics mine.
9. *Ibid.*, p. 1.
10. See W. Heisenberg, "Quantum Theory and Its Interpretation", p. 105.
11. Oscar Klein, "Glimpses of Niels Bohr as Scientist and Thinker", in S.Rozental (ed.), *Niels Bohr*, 74–93, cf. p. 85.
12. N. Bohr, H.A. Kramers and J.A. Slater, "The Quantum Theory of Radiation", *Philosophical Magazine*, 47, 1924, 785–802, reprinted in *Collected Works*, Vol. 5, 101–118. See, further, K. Stolzenburg's "Introduction" to this paper in the same volume; H. Folse, *The Philosophy of Niels Bohr*, pp 72 ff.; and D. Murdoch, *Niels*

Bohr's philosophy of physics, pp. 23–9.
13. "On the Quantum Theory of Line-Spectra", p. 7.
14. Max Jammer, *The Conceptual Development of Quantum Mechanics*, pp. 113–14.
15. "On the Application of the Quantum Theory to Atomic Structure", p. 21.
16. This is evident from a letter of 21th April 1925 from Bohr to Geiger, *Collected Works*, Vol. 5, p. 253; and from a letter of 1st May 1925 from Bohr to Max Born, *Ibid.*, p. 311. See further D. Murdoch, *Niels Bohr's philosophy of physics*, pp. 29 ff.
17. "Atomic Theory and Mechanics" (1925), *ATDN*, p. 34, as well as *Collected Works*, Vol. 5, p. 276.
18. *Collected Works*, Vol. 5, p. 274.
19. *Idem.*
20. Letter from Schrödinger to Bohr, 23th October 1926, on his return to Zurich after his visit to Bohr in Copenhagen. Quoted from *Collected Works*, Vol. 6, p. 12.
21. See *Oversigt over Det kgl. Danske Videnskabernes Selskabs Forhandlinger 1926–27* (The bulletin of the Royal Academy of Sciences and Letters), p. 28–29.
22. Letter to Meyerson 30th December 1926, in *Correspondance entre Harald Høffding and Emile Meyerson*, p. 131.
23. See W. Heisenberg: "The Development of the Interpretation of the Quantum Theory", in W.Pauli (ed.), *Niels Bohr and the Development of Physics*, London 1955, p. 15.

CHAPTER VI

1. See N. Bohr, "Introductory Survey" (1929), in *ATDN*, 1–24, p. 1 and p. 18.
2. N. Bohr, "The Quantum Postulate and the Recent Development of Atomic Theory" (1928), in *ATDN*, 52–91, p. 53.
3. N. Bohr, "The Quantum of Action and the Description of Nature" (1929), in *ATDN*, 92–101, p. 94.
4. N. Bohr, "The Rutherford Memorial Lectures 1958" (1961), in his *Essays 1958–1962 on Atomic Physics and Human Knowledge* [*Essays*], 30–73, J. Wiley & Sons, London 1963, p. 59.
5. *ATDN*, p. 16.
6. *Ibid.*, p. 93.
7. N. Bohr, "The Atomic Theory and the Fundamental Principles Underlying the Description of Nature" (1929), in *ATDN*, 102–19, pp. 116–17. The phrase "modes of perception" is a rendering of the Danish word "anskuelsesformer".
8. *Psykologi*, p. 263.
9. This has been noticed independently by Murdoch in *Niels Bohr's philosophy of physics*, p. 72
10. *ATDN*, p. 8.
11. N. Bohr, "Causality and Complementarity" (1936), *Philosophy of Science*, 4, 1937, 289–98, p. 293.
12. *Psykologi*, p. 271.
13. *ATDN*, p. 5.
14. *Psykologi*, pp. 262–63.
15. *Ibid.*, p. 271.
16. *ATDN*, p. 93.
17. *Ibid.*, p. 100.
18. *Ibid.*, p. 115.
19. *Ibid.*, p. 104.
20. *Ibid.*, p. 93.
21. *Ibid.*, p. 91.
22. *Ibid.*, p. 10.
23. *Ibid.*, p. 93.

24. *Ibid.*, p. 96.
25. *Ibid.*, p. 115; cf. also pp. 53–54.
26. *Ibid.*, p. 68; cf also pp. 11–12.
27. *Ibid.*, p. 17.
28. N. Bohr, "Natural Philosophy and Human Cultures" (1938), in *APHK*, 23–31, pp. 25–26.
29. *Ibid.*, p. 26.
30. N. Bohr, "Quantum Physics and Philosophy – Causality and Complementarity" (1958), in *Essays*, 1–7, pp. 3–4.
31. *ATDN*, p. 55.
32. See *ibid.*, p. 11.
33. *Ibid.*, p. 10.
34. See *ibid.*, p. 62 and p. 73.
35. *Ibid.*, p. 56.
36. *Ibid.*, pp. 55–56.
37. *Idem.*
38. See D. Murdoch, *Niels Bohr's philosophy of physics*, p. 67.
39. In my book *The reality of the future*, Odense University Press 1989, pp. 300 ff., I have, for instance, suggested that the wave-particle duality might be explained away on an assumption of the existence of advanced and superluminal particles operating under the limits of the Heisenberg uncertainty relations.
40. *ATDN*, pp. 68–69.
41. *Ibid.*, p. 111.
42. "Causality and Complementarity", p. 290.
43. N. Bohr, "Discussion with Einstein on Epistemological Problems in Atomic Physics", p. 211, in P.A. Schilpp (ed.), *Albert Einstein: Philosopher-Scientist.* Library of Living Philosophers, Northwestern University Press, Evanston 1949, 199–242, Reprinted in *APHK*, 32–66.
44. See also *Essays*, p. 6. Here Bohr likewise talks about the renunciation of the ideal of causality.
45. N. Bohr, "Light and Life" (1933), in *APHK*, 3–12, p. 7.
46. *ATDN*, p. 117.
47. Letter to Pauli, 1 July 1929, *BSC*: 14. See *Collected Works*, vol 6, p. 443.
48. *ATDN*, p. 96.
49. *Ibid.*, p. 116, italics mine.
50. *Ibid.*, p. 15.
51. *APHK*, p. 52.
52. *Filosofiske Problemer*, pp. 48 ff.
53. *ATDN*, p. 100.
54. H. Høffding, *Excerpts and notices*, 287 Quarto R. pp. 264–67. The Royal Library, N.Kgl.Saml.3353. Mss. V–2. Høffding's own reference to his *Psykologi* is to the 10th edition 1925 pp. 13–15.
55. *ATDN*, p. 96.
56. This objection has been raised by Henry Folse in a private communication.
57. N. Bohr, "Biology and Atomic Physics" (1937), in *APHK*, 13–22, pp. 21–22, italics mine.
58. *APHK*, p. 11.
59. *Den menneskelige Tanke*, p. 14 ff.
60. *ATDN*, p. 99.
61. *Ibid.*, p. 96.
62. *Ibid.*, p. 99.
63. See L. Rosenfeld, "Niels Bohr in the Thirties", p. 121.
64. L. Rosenfeld, "Niels Bohr's Contributions to Epistemology", *Physics Today*, 16 (1963), p. 48.

65. N. Bohr, "The Unity of Knowledge" (1960), in *Essays*, 8–16, p. 13.
66. *Idem.*
67. *ATDN*, p. 100, italics mine.
68. *Idem.*
69. *APHK*, p. 11.
70. This objection has been raised by David Favrholdt in conversation.
71. See *Psykologi*, pp. 88–89, note 1; and *Filosofiske Problemer*, p. 24–25.
72. *ATDN*, pp. 116–17.
73. *APHK*, p. 22.
74. *Ibid.*, p. 96.
75. N. Bohr, "Light and Life", in *Nature*, 131 (1933), p. 422. A somewhat different version appears in *APHK*, 3–12.
76. *APHK*, p. 21.
77. *Ibid.*, p. 8.
78. *Ibid.*, p. 10.
79. *Ibid.*, p. 21.
80. *Idem.*
81. See H. Folse, *The Philosophy of Niels Bohr*, pp. 185 ff.
82. *Essays*, p. 102.
83. *APHK*, p. 7.
84. *Ibid.*, p. 20.
85. *Ibid.*, pp. 20–21.
86. This is also what Henry Folse has argued. See his "Complementarity and the Description of Nature in Biological Science", in *Biology and Philosophy*, 5 (1990), 211–24.

CHAPTER VII

1. *ATDN*, p. 115.
2. See, for instance, K. Popper, "Quantum Mechanics Without 'the Observer'" in M. Bunge (ed.), *Quantum Theory and Reality*, New York 1967, 7–44; as well as M. Bunge's introduction "The Turn of the Tide". See also M. Bunge, "Strife about Complementarity", *Brit. J. Phil. Sci.*, 6, 1955, 1–12, and 141–54.
3. *ATDN*, p. 116.
4. *Ibid.*, p. 1.
5. *Ibid.*, p. 97.
6. D. Murdoch, *Niels Bohr's philosophy of physics*, p. 139 ff.
7. *ATDN*, p. 18.
8. See A. Grünbaum, "Complementarity in Quantum Physics and its Philosophical Generalization", in *Journal of Philosophy*, 54, 1957; and P. Feyerabend, "Problems of Microphysics" in R.G.Colodny (ed.), *Frontiers of Science and Philosophy*, Pittsburgh 1962, pp. 196 ff; as well as P. Feyerabend, "On a Recent Critique of Complementarity: Part I", in *Journal of Philosophy*, 35, 1968, 309–31, and "Part II", in *Journal of Philosophy*, 36, 1969, 82–105, esp. pp. 94 ff. See also D. Murdoch's criticism, *Niels Bohr's philosophy of physics*, pp. 140 ff.
9. *ATDN*, p. 115.
10. *Ibid.*, p. 56.
11. *Ibid.*, p. 56.
12. *APHK*, p. 26.
13. *Ibid.*, p. 25.

14. *ATDN*, p. 66.
15. See *ibid.*, pp. 62 ff. and p. 77.
16. See L. Rosenfeld, "Misunderstandings about the foundation of quantum theory", in S.Körner (ed.), *Observation and Interpretation*, London 1957, 41–45, p. 42.
17. P.K. Feyerabend, "On a Recent Critique of Complementarity: Part I", p. 322. See also his "Problems of Microphysics", p. 217 ff.
18. See *ATDN*, p. 93 and p. 103.
19. *Ibid.*, p. 19
20. See, for instance, A. Fine, *The Shaky Game: Einstein, Realism and the Quantum Theory*, The University of Chicago Press, 1986, pp. 34–35.
21. For a historical account of the Bohr-Einstein debate, see M. Jammer, *The Philosophy of Quantum Mechanics*, John Wiley & Sons, New York 1974, ch. 5.
22. A. Einstein, B. Podolsky and N. Rosen, "Can Quantum-Mechanical Description of Physical Reality Be Considered Complete?", in *Physical Review*, 47, 1935, 777–80.
23. For a very penetrating analysis of the EPR argument and the Bohrian response, see C.A. Hooker, "The Nature of Quantum Mechanical Reality: Einstein Versus Bohr", in R.G. Colodny (ed.), *Paradigms and Paradoxes: The Philosophical Challenge of the Quantum Domain*, Pittsburgh 1972, pp. 67 - 302.
24. D. Howard, "Einstein on Locality and Separability", in *Stud. Hist. Phil. Sci.*, 16, 1985, 171–201. See also Murdoch's comments on Howard in his *Niels Bohr's philosophy of physics*, p. 173 f.
25. *Ibid*, p. 179.
26. L. Rosenfeld, "Bohr in the Thirties", pp. 128–29.
27. N. Bohr, "Quantum Mechanics and Physical Reality", in *Nature*, 136, 1935,
28. N. Bohr, "Can Quantum-Mechanical Description of Physical Reality be Considered Complete?", in *Physical Review*, 48, 1935, 696–702.
29. Bohr's reply in *Physical Review* was received by the editor on July 13, the very day its appearance was announced in the letter in *Nature*.
30. N. Bohr, "Can Quantum Mechanical Description of Physical Reality be Considered Complete?", p. 697.
31. *Ibid.*, p. 700
32. C.A. Hooker, "The Nature of Quantum Mechanical Reality: Einstein Versus Bohr", pp. 223–24.
33. See J.P. Jarrett, "On the Physical Significance of the Locality Conditions in the Bell Arguments", in *Noûs*, 18 (1984), 569–89.
34. I prefer Don Howard's way of putting these two conditions rather than Jon Jarrett's for various reasons. See D. Howard, "Einstein on Locality and Separability", p. 196. One of them is that his formulation is more suitable for grasping in what direction a Bohrian response to the work of Bell would have to go.
35. N. Bohr, "The Causality Problem in Atomic Physics", in *New Theories in Physics*, International Institute of Intellectual Collaboration, Paris 1939, 11–30, p. 21.
36. I have argued for this point of view in J. Faye, *The reality of the future*, pp. 300 f.
37. A. Einstein, "Reply to Criticism", in P.A. Schilpp (ed.), *Albert Einstein: Philosopher-Scientist*, p. 681.
38. This seems to be true of H. Folse in *The Philosophy of Niels Bohr*, p. 154; and of J. Honnor in *The Description of Nature*, p. 79. However, D. Murdoch is more aware of changes in Bohr's arguments in *Niels Bohr's philosophy of physics*, p. 145.
39. See, for instance, A. Fine, *The Shaky Game: Einstein, Realism and the Quantum Theory*, p. 35.
40. "Causality and Complementary", p. 293.
41. "The Causality Problem in Atomic Physics", p. 19.
42. "Can Quantum-Mechanical Description of Physical Reality be Considered Complete?", p. 699.
43. *Essays*, p. 5.

44. *ATND*, p. 57, 63, 65, 66, 67, 68, 80, and p. 114.
45. *Ibid.*, p. 11.
46. See *APHK*, p. 39, 40, 43, 44, and p. 72; *Essays*, p. 5, 78, and p. 91; as well as "Causality and Complementarity", p. 292, with the 1938 paper "The Causality Problem in Atomic Physics", p. 18, 20, 21, and p. 23 as the only exception.
47. *APHK*, p. 50. See also *Essays*, p. 3.
48. *Ibid.*, p. 39.
49. *Idem.*
50. *Essays*, p. 1.
51. *Ibid.*, p. 59.
52. See *APHK*, pp. 80–81.
53. *Ibid.*, pp. 40–41.
54. *Ibid.*, p. 50.
55. *Ibid.*, p. 228.
56. *Ibid.*, p. 210.
57. See *APHK*, p. 64 and p. 72; as well as *Essays*, p. 5.
58. See *APHK*, p. 25, 26, 30, 61, 72, and p. 89; as well as *Essays*, p. 4.
59. *Essays*, p. 6.
60. *Ibid.*, p. 238.
61. "The Causality Problem in Atomic Physics", p. 20.
62. See *ATDN*, p. 11, 65, 68, and p. 115; as well as *APHK*, pp. 6–7.
63. See H. Folse, *The Philosophy of Niels Bohr*, pp. 156 f.
64. "The Causality Problem in Atomic Physics", pp. 21–22.
65. *APHK*, p. 72, and p. 74.
66. *Essays*, p. 25.
67. D. Murdoch, *Niels Bohr's philosophy of physics*, p. 148
68. *Ibid.*, pp. 149 ff.
69. Since writing my review of Murdoch's *Niels Bohr's philosophy of physics*, in *Isis*, 81, 2 (1990), 278–79 I have become more critical of his interpretation of Bohr's conception of physical properties and the conditions under which we may meaningfully ascribe such to atomic objects.

CHAPTER VIII

1. *Essays*, p. 5.
2. J. Faye, *The reality of the future*, pp. 85 ff.
3. See C. Wright, *Realism, Meaning and Truth*, Oxford 1987, p. 5 and pp. 148–49.
4. S.A. Pedersen, "Kan matematiske og fysiske teorier tolkes realistisk?" (May mathematical and physical theories be interpreted realistically?), *Nyere dansk filosofi*, Philosophia 1984, 1–19.
5. W.H. Newton-Smith, *The Rationality of Science*, Routledge & Kegan Paul, London 1981, p. 40 f.
6. H. Folse, *The Philosophy of Niels Bohr*, p. 237.
7. *Ibid.*, pp. 234–37.
8. *Ibid.*, pp. 231–32 and 243–45.
9. *APHK.*, p. 26.
10. H. Folse, *The Philosophy of Niels Bohr*, pp. 237–40 and 243–45.
11. *Ibid.*, p. 240.
12. *Idem.*
13. *Ibid.*, p. 238; cf. p. 257 also.
14. See some of Folse's more recent papers, "Niels Bohr, Complementarity, and

Realism", *PSA 1986: Proceedings of the Biennial Meeting of the Philosophy of Science Association*, Vol. I, ed. by A. Fine and P. Machamer, East Lansing, PSA, 1986, 96–104; "Niels Bohr's Concept of Reality", *Symposion on the Foundations of Modern Physics 1987: The Copenhagen Interpretation 60 Years After the Come Lecture - Joensuu, Finland, August 6–8, 1987*, by Pekka Lahti and Peter Mittelstaedt, World Scientific Publishing, Singapore 1987, 161–79; and "Complementarity and our Knowledge of Nature", *Nature, Cognition, and System*, Vol 2, ed. by Mark Carvallo, Kluwer, Dordrecht 1990.
15. *ATDN*, pp. 102-03, italics mine.
16. *Ibid.*, p. 93.
17. E. Mach, *Die Mechanik in Ihrer Entwicklungsgeschichte*, Darmstadt 1976, pp. 466–67.
18. E. Mach, *Erkenntnis und Irrtum*, Darmstadt 1976, pp. 8 ff.
19. H. Folse, *The Philosophy of Niels Bohr*, pp. 247 ff.
20. Niels Bohr Archive, *BSC*: 27. Letter of 2 March 1953 from Niels Bohr to Max Born in which Bohr quotes Born from a preprint of the article "The Interpretation of Quantum Mechanics".
21. Niels Bohr Archive, *BSC*: 25. Letter of 2. March 1953 from Niels Bohr to Max Born.
22. Niels Bohr Archive, *BSC*: 27. Letter of 10 March 1953 from Max Born to Niels Bohr.
23. Niels Bohr Archive, *BSC*: 27. Letter of 26 March 1953 from Niels Bohr to Max Born.
24. H. Høffding, "Om Realisme i Videnskab og Tro" (On Realism in Science and Faith), *Mindre Arbejder*, I, p. 7.
25. *Den menneskelige Tanke*, p. 106.
26. *Idem.*
27. *APHK*, pp. 78–79.
28. *ATDN*, p. 18.
29. *Essays*, p. 7; cf. p. 3. also.
30. See M. Dummett, "Realism", *Synthese*, 52 (1982), 55–112, pp. 94 ff.
31. *APHK*, p. 67.
32. *Essays*, p. 1.
33. *Ibid.*, p. 4.
34. *APHK*, p. 72; cf. also *Essays*, p. 3.
35. *APHK*, p. 80.
36. *Ibid.*, p. 26.
37. See, for instance, M. Dummett, "The Philosophical Basis of Intuitionistic Logic", *Truth and other Enigmas*, Duckworth, London 1978, 215–247.
38. *APHK*, p. 71.
39. *Ibid.*, p. 67.
40. *Ibid.*, p. 68
41. See, for instance, W.V.O. Quine, *From a Logical Point of View*, Cambridge, Mass. 1964, pp. 16 f. and p. 44
42. Aa. Petersen, *Quantum Physics and the Philosophical Tradition*, The M.I.T. Press, Cambridge 1968, pp. 135–37.
43. *Ibid.*, p. 138.
44. *Idem.*
45. *Ibid.*, p. 147 f.
46. *Ibid.*, pp. 163–64
47. *Ibid.*, 149.
48. *Ibid.*, pp. 185 ff.
49. *APHK*, p. 90.
50. *Essays*, p. 6.
51. See, for instance, *APHK*, p. 73.
52. "The Causality Problem in Atomic Physics", p. 19; see also pp. 23–24.
53. D. Murdoch, *Niels Bohr's philosophy of physics*, p. 107.

54. *Ibid*, p. 151 and p. 230.
55. *APHK*, p. 51.
56. D. Murdoch, *Niels Bohr's philosophy of physics*, p. 192.

EPILOGUE

1. *APHK*, p. 88.

Bibliography

Blædel, N., *Harmoni og Enhed, Niels Bohr, en biografi*, Rhodos, Copenhagen 1985.

Bohr, N., "On the Quantum Theory of Line-Spectra" (1918), *Det kgl. Danske Vid. Selsk. Skrifter. Naturvidenskabelig og Matematisk Afdeling*, 8. række, IV, Copenhagen 1918–1922.

—, "On the Series Spectra of Elements" (1920), printed in N. Bohr, *The Theory of Spectra and Atomic Constitution*, Cambridge 1922, 20–60.

—, "On the Application of the Quantum Theory to Atomic Structure" (1922), *Proceedings of the Cambridge Philosophical Society Supplement*, 22, 1924, 1–42.

—, "Atomteori og Bølgemekanik", report of a lecture printed in *Oversigt over Det kgl. Danske Vid. Selsk. Forhandlinger 1926–1927.*

—, "The Quantum Postulate and the Recent Development of Atomic Theory" (1927), *Atti del Congresso Internazionale dei Fisici 11–20 Settembre 1927*, Como-Pavia-Roma, Volume Secondo, Nicola Zanichelli, Bologna 1928, 565–88.

—, "The Quantum Postulate and the Recent Development of Atomic Theory" (1928), *Nature*, 121, 1928, 580–90, reprinted in *ATDN*, 52–91.

—, "Ved Harald Høffdings 85-årsdag" (1928), *Berlingske Tidende*, 10 March 1928. Evening edition.

—, "The Quantum of Action and the Description of Nature" (1929), in *ATDN*, 92–101.

—, "The Atomic Theory and the Fundamental Principles Underlying the Description of Nature" (1929), in *ATDN*, 102–19.

—, "Introductory Survey" (1929), in *ATDN*, 1–24.

—, "Mindeord over Harald Høffding" (1931), *Oversigt over Det kgl. Danske Vid. Selsk. Forhandlinger 1931–1932*, 131–36.

—, "Light and Life" (1932), *Nature*, 131, 1933, 423–59.

—, "Light and Life" (1933), in *APHK*, 3–12.

—, *Atomic Theory and the Description of Nature* [abbr. *ATDN*], Cambridge University Press, Cambridge 1934.

—, "Quantum Mechanics and Physical Reality" (1935), *Nature*, 136, 1935, 65.

—, "Can Quantum-Mechanical Description of Physical Reality be Considered Complete?" (1935), *Physical Review*, 48, 1935, 696–702.

—, "Causality and Complementarity" (1936), *Journal of Philosophy*, 4, 1937, 289–98.

—, "Biology and Atomic Physics" (1937), in *APHK*, 13–22.

—, "The Causality Problem in Atomic Physics" (1938), in *New Theories in Physics*, International Institute of Intellectual Collaboration, Paris 1939, 11–30.

—, "Natural Philosophy and Human Cultures" (1938), in *APHK*, 23–31.

—, "Discussion with Einstein on Epistemological Problems in Atomic Physics" (1949), in P.A. Schilpp (ed.), *Albert Einstein: Philosopher-Scientist*, Library of Living Philosophers, Northwestern University Press, Evanston 1949, 199–242, reprinted in *APHK*, 32–66.

——, "Unity of Knowledge" (1954), in *APHK*, 67–82.
——, "Atoms and Human Knowledge" (1956), in *APHK*, 83–93.
——, "Physical Science and the Problem of Life" (1949 and 1957), in *APHK*, 94–101.
——, *Atomic Physics and Human Knowledge* [abbr. *APHK*], J. Wiley & Sons, New York 1958.
——, "Quantum Physics and Philosophy - Causality and Complementarity" (1958), *Philosophy in the Mid-century*, La Nuova Italia Editrice, Florence 1958, reprinted in *Essays*, 1–7.
——, "The Unity of Human Knowledge" (1960), in *Essays*, 8–16.
——, "The Rutherford Memorial Lectures 1958" (1961), in *Essays*, 70–73.
——, *Essays 1958–1962 on Atomic Physics and Human Knowledge* [abbr. *Essays*], J. Wiley & Sons, New York 1963.
——, *Niels Bohr: Collected Works*, (eds.) L. Rosenfeld, J. Rud Nielsen, E. Rüdinger *et al.*, vol 1-, North-Holland Publishing Company, Amsterdam 1971-.
——, Kramers H.A. and Slater, J.A., "The Quantum Theory of Radiation", *Philosophical Magazine*, 47, 1924, 785–802.
Brandt, F., "Læreren og Humanisten", in E. Rubin (ed.), *Høffding in Memoriam*, 31–39.
—— *et al.* (eds.), *Correspondance entre Harald Høffding et Emile Meyerson*, Copenhagen 1939.
Bunge, M., "Strife about Complementarity", *British Journal for Philosophy of Science*, 6, 1955, 1–12, and 141–54.
—— (ed.), *Quantum Theory and Reality*, Springer Verlag, New York 1967.
——, "The Turn of the Tide", in M. Bunge (ed.) *Quantum Theory and Reality*, 1–12.
Cohn, G., "Harald Høffding og hans Filosofi", *Tilskueren*, 50, 1933, 103–17.
Dummett, M., "The Philosophical Basis of Intuitionistic Logic", in his *Truth and other Enigmas*, 215–47, Duckworth, London 1978.
——, "Realism", *Synthese*, 52, 1982, 55–112.
Einstein, A., Podolsky B. and Rosen, N., "Can Quantum-Mechanical Description of Physical Reality be Considered Complete?", *Physical Review*, 47, 1935, 777–80.
Einstein, A., "Reply to Criticism", in P.A. Schilpp (ed.), *Albert Einstein: Philosopher-Scientist*, Library of Living Philosophers, Northwestern University Press, Evanston 1949.
Favrholdt, D., "Bevidsthedsproblemet i Harald Høffding's filosofi", *Det kgl. Danske Vid. Selsk. Historiske-filosofiske Meddelelser*, 44, no. 4, 1969, 1–36.
——, "Niels Bohr and Danish Philosophy", *Danish Yearbook of Philosophy*, 13, 1976, 206–20.
——, "On Høffding and Bohr. A reply to Jan Faye", *Danish Yearbook of Philosophy*, 16, 1979, 73–77.
——, "The Cultural Background of the Young Bohr", *Rivistra di Storia della Scienza*, 2, no. 3, 1985, 445–61.
Faye, J., "The Influence of Harald Høffding's Philosophy on Niels Bohr's Interpretation of Quantum Mechanics", *Danish Yearbook of Philosophy*, 16, 1979, 37–72.
——, "The Bohr-Høffding Relationship Reconsidered", *Studies in History and Philosophy of Science*, 19, no. 3, 1988, 321–46.
——, *The reality of the future. An essay on time, causation and backward causation*, Odense University Press, Odense 1989.
——, "Review of Dugald Murdoch's 'Niels Bohr's philosophy of physics'", *Isis*, 81, no. 2, 1990, 278–79.
Feyerabend, P., "Problems of Microphysics", in R.G. Colodny (ed.), *Frontiers of Science and Philosophy*, University of Pittsburgh Series in Philosophy of Science, vol. 1, University of Pittsburgh Press, Pittsburgh 1962.
——, "On Recent Critique of Complementarity: Part I", *Journal of Philosophy*, 35, 1968, 309–31.
——, "On Recent Critique of Complementarity: Part II", *Journal of Philosophy*, 36, 1969, 82–105.

Fine, A., *The Shaky Game: Einstein, Realism and the Quantum Theory*, The University of Chicago Press, Chicago 1986.

Folse, H., *The Philosophy of Niels Bohr. The Framework of Complementarity*, North-Holland Publishing Company, Amsterdam 1985.

—, "Niels Bohr, Complementarity, and Realism", in A. Fine and P. Machamer (eds.), *PSA 1986. Proceedings of the Biennial Meeting of the Philosophy of Science Association*, East Lansing. PSA 1986, 96–104.

—, "Niels Bohr's Concept of Reality, in P. Lahti and P. Mittelstaedt (eds.), *Symposion on the Foundations of Modern Physics 1987: The Copenhagen Interpretation 60 Years After the Como Lecture - Joensuu, Finland, August 6–8, 1987*, World Scientific Publishing, Singapore 1987, 161–79.

—, "Complementarity and the Description of Nature in Biological Science", *Biology and Philosophy*, 5, 1990, 211–24.

—, "Complementarity and our Knowledge of Nature", in M. Carvallo, (ed.), *Nature, Cognition, and System*, vol. 2, Kluwer Academic Publishers, Dordrecht 1990.

Fuglsang-Damgaard, H., "Harald Høffding 1843–1943", *Dansk Teologisk Tidsskrift*, 6, 1943, 225–43.

Grünbaum, A., "Complementarity in Quantum Physics and its Philosophical Generalization", *Journal of Philosophy*, 54, 1957, 713–27.

Heisenberg, W., "The Development of the Interpretation of the Quantum Theory", in W. Pauli (ed.), *Niels Bohr and the Development of Physics*, 12–29.

—, "Quantum Theory and Its Interpretation", in S. Rozental (ed.), *Niels Bohr: His life and work as seen by friends and colleagues*, 94–108.

Holton, G., "The Roots of Complementarity", *Daedalus*, 99, 1970, 1015–55.

Honnor, J., *The Description of Nature. Niels Bohr and the Philosophy of Quantum Physics*, Clarendon Press, Oxford 1987.

Hooker, C.A., "The Nature of Quantum Mechanical Reality: Einstein Versus Bohr", in R.G. Colodny (ed.), *Paradigms and Paradoxes: The Philosophical Challenge of the Quantum Domain*, 67–302, University of Pittsburgh Series in the Philosophy of Science, vol. 5, University of Pittsburgh Press, Pittsburgh 1972.

Howard, D., "Einstein on Locality and Separability", *Studies in History and Philosophy of Science*, 16, 1985, 171–201.

Høffding, H., *Psykologi i Omrids på Grundlag af Erfaringen*, Copenhagen 1882 and later enlarged and revised editions.

—, "Om Realisme i Videnskab og Tro" (1884), in *Mindre Arbejder I*, 1–14.

—, *Etik. En Fremstilling af de etiske Principper og deres Anvendelse paa de vigtigste Livsforhold*, Copenhagen 1887.

—, *Søren Kierkegaard som Filosof*, Copenhagen 1892 and 1919.

—, *Den nyere Filosofis Historie. En fremstilling af Filosofiens Historie fra Renaissancens Slutning til vore Dage*, vol. I-II, Copenhagen 1894–1895.

—, *A History of Modern Philosophy* (The English translation of *Den nyere Filosofis Historie*), vol. I-II, Dover, New York 1955.

—, "Det psykologiske Grundlag for logiske Domme", *Det kgl. Danske Vid. Selsk. Skrifter*, 6. række, historisk og filosofisk afd. I, bd. 4, Copenhagen 1893–1899, 343–403.

—, "Om Vitalisme" (1898), in *Mindre Arbejder I*, 40–50.

—, *Filosofiske Problemer*, Københavns Universitets Festskrift, Copenhagen 1902.

—, *Formel Logik*, 4th ed., Copenhagen 1903

—, *Moderne Filosofer*, Copenhagen 1904.

—, "Begrebet Analogi og dets filosofiske Betydning" (1904), in *Mindre Arbejder II*, 33–46.

—, "En filosofisk Bekendelse" (1904), in *Mindre Arbejder III*, 18–27.

—, *Mindre Arbejder I-II*, Copenhagen 1905.

—, "Begrebet Villie" (1905 og 1912), in *Mindre Arbejder III*, 28–52.

—, "On analogy and its philosophical importance", *Mind*, 14, 1905, 199–209.

—, *Formel Logik*, 5th ed., Copenhagen 1907.

——, "Charles Darwin og Filosofien" (1909), in *Mindre Arbejder III*, 202- 28.
——, *Den menneskelige Tanke, dens Former og dens Opgave*, Copenhagen 1910.
——, "Mindetale over Christian Bohr", *Tilskueren*, 1911, 209–12.
——, *Mindre Arbejder III*, Copenhagen 1913.
——, "Totalitet som Kategori", *Det kgl. Danske Vid. Selsk. Skrifter*, 6. række, historisk-filosofisk afd. II, Copenhagen 1917.
——, "Spinozas Ethica", *Det kgl. Danske Vid. Selsk. Skrifter*, 7. række, historisk-filosofisk afd. II, vol. 3, Copenhagen 1918.
——, "Relation som Kategori", *Det kgl. Danske Vid. Selsk. Filosofiske Meddelelser I*, no. 3, Copenhagen 1921.
——, "Begrebet Analogi", *Det kgl. Danske Vid. Selsk. Filosofiske Meddelelser I*, no. 4, Copenhagen 1923.
——, "Erkendelsesteori og Livsopfattelse", *Det kgl. Danske Vid. Selsk. Filosofiske Meddelelser II*, no. 1, Copenhagen 1925.
——, *Erindringer*, Copenhagen 1928.
——, "Bemærkninger om Erkendelsesteoriens nuværende stilling", *Det kgl. Danske Vid. Selsk. Filosofiske Meddelelser II*, no. 2, Copenhagen 1930.
——, "Psykologi og Autobiografi", *Det kgl. Danske Vid. Selsk. Filosofiske Meddelelser II*, no. 3, Copenhagen 1943.
James, W., *The Principle of Psychology I-II*, London 1891.
Jammer, M., *The Conceptual Development of Quantum Mechanics*, McGraw-Hill, New York 1966.
——, *The Philosophy of Quantum Mechanics*, J. Wiley & Sons, New York 1974.
Jarrett, J.P., "On the Physical Significance of the Locality Conditions in the Bell Arguments", *Noûs*, 18, 1984, 569–89.
Kalckar, J., "General introduction to volume 6 and 7", in *Niels Bohr: Collected Works*, vol. 6, xvii-xxvi.
Klein, O., "Glimpses of Niels Bohr as Scientist and Thinker", in S. Rozental (ed.), *Niels Bohr: His life and work as seen by friends and colleagues*, 74–93.
Lübcke, P., "F.C. Sibbern: Epistemology as Ontology", *Danish Yearbook of Philosophy*, 13, 1976, 84–104.
Mach, E., *Die Mechanik in Ihrer Entwicklungsgeschichte*, Darmstadt 1976.
——, *Erkenntnis und Irrtum*, Darmstadt 1976.
Maxwell, J.C., *The Scientific Papers*, vol. I-II, Cambridge 1890.
Murdoch, D., *Niels Bohr's philosophy of physics*, Cambridge University Press, Cambridge 1987.
Newton-Smith, W.H., *The Rationality of Science*, Routledge & Kegan Paul, London 1981.
Nielsen, J. Rud, "Memories of Niels Bohr", *Physics Today*, 16, no. 10, 1963, 22–30.
Nielsen, R., *Almindelig Videnskabslære i Grundtræk*, Copenhagen 1880.
Pauli, W. (ed.), *Niels Bohr and the Development of Physics*, Pergamon Press, London 1955.
Petersen, Aa., "The Philosophy of Niels Bohr", *Bulletin of the Atomic Scientist*, 19, no. 7, 1963, 8–14.
——, *Quantum Physics and the Philosophical Tradition*, The M.I.T. Press, Cambridge, Mass. 1968.
Petersen, S.A., "Kan matematiske og fysiske teorier tolkes realistisk", *Nyere dansk filosofi*, Philosophia, Aarhus 1985, 1–19.
Popper, K., "Quantum Mechanics Without 'the Observer'" in Bunge, M. (ed.), *Quantum Theory and Reality*, 7–44.
Quine, W.V.O., *From a Logical Point of View*, Cambridge, Mass. 1964.
Rosenfeld, L., "Misunderstandings about the foundation og quantum theory", in S. Körner (ed.), *Observation and Interpretation*, 41–45.
——, "Niels Bohr's Contributions to Epistemology", *Physics Today*, 16, no. 10, 1963, 47–54.
——, "Niels Bohr in the Thirties", in S. Rozental (ed.), *Niels Bohr: His life and work as seen by friends and colleagues*, 114–36.

Rozental, S. (ed.), *Niels Bohr: His life and work as seen by friends and colleagues*, North-Holland Publishing Company, Amsterdam 1967.
Rubin, E. (ed.), *Harald Høffding in Memoriam. Fire taler holdt på Københavns Universitet paa Harald Høffdings 89 Aars Dag 11. marts 1932*, Copenhagen 1932.
Skov, P., *Aarenes Høst. Erindringer fra mange Lande i urolige Tider*, Copenhagen 1961.
Slomann, V., "Minder om samvær med Niels Bohr", *Politiken*, 7.10.1955.
Stolzenburg, K., "Introduction", in *Niels Bohr: Collected Works*, vol. 5, 3–96.
Stybe, S.E., "Niels Treschow (1751–1833). A Danish Neoplatonist", *Danish Yearbook of Philosophy*, 13, 1976, 29–47.
Witt-Hansen, J., "Leibniz, Høffding and the "Ekliptika" Circle", *Danish Yearbook of Philosophy*, 17, 1980, 31–58.
Wright, C., *Realism, Meaning and Truth*, Oxford 1987.

[illegible]sen, S. (ed.), Niel[illegible] [illegible] and [illegible] [illegible]
[illegible] Publishing [illegible]
[illegible] Universitet [illegible]
[illegible] København, 1977 [illegible]
[illegible] P. [illegible] København [illegible]
[illegible] Niels Bohr [illegible] 1985.
[illegible] K. [illegible] Niels Bohr [illegible] Facts [illegible]
[illegible] Danish Yearbook of Philosophy [illegible]
[illegible] Complementarity and the Copenhagen [illegible]
[illegible] (1980), 31 [illegible]
[illegible] (1985) [illegible]

Index

Science and Philosophy

Series Editor:

Nancy J. Nersessian, *Program in History of Science, Princeton University*

1. N.J. Nersessian: *Faraday to Einstein: Constructing Meaning in Scientific Theories.* 1984 ISBN Hb 90-247-2997-1 / Pb (1990) 0-7923-0950-2
2. W. Bechtel (ed.): *Integrating Scientific Disciplines.* 1986 ISBN 90-247-3242-5
3. N.J. Nersessian (ed.): *The Process of Science.* Contemporary Philosophical Approaches to Understanding Scientific Practice. 1987 ISBN 90-247-3425-8
4. K. Gavroglu and Y. Goudaroulis: *Methodological Aspects of the Development of Low Temperature Physics 1881–1956.* Concepts out of Context(s). 1989 ISBN 90-247-3699-4
5. D. Gooding: *Experiment and the Making of Meaning.* Human Agency in Scientific Observation and Experiment. 1990 ISBN 0-7923-0719-4
6. J. Faye: *Niels Bohr: His Heritage and Legacy.* An Anti-realist View of Quantum Mechanics. 1991 ISBN 0-7923-1294-5

Kluwer Academic Publishers – Dordrecht / Boston / London

9 780792 312949